GLOBALISATION

access to geography

GLOBALISATION

Paul Guinness

Hodder & Stoughton

A MEMBER OF THE HODDER HEADLINE GROUP

Acknowledgements
The publishers would like to thank the following individuals, institutions and companies for permission to reproduce copyright illustrations in this book:

Sarah Bosely © *The Guardian*, page 93; P Cramp © Anforme, page 21; Charlotte Denny © *The Guardian*, page 32; © *The Financial Times* (25/09/02), page 84; Chris Guinness, page 58; Mary Guinness, page 25; Robert Jones, WTO – Recent Developments, *Oxford Review of Economic Policy*, 2000, Volume 29, Issue 2, page 35 © Oxford University Press, page 34; © Kommunalverband Ruhrgebiet, page 73; © Telegraph Group Limited (2003), page 97 © UNCTAD, page 29.

The publishers would also like to thank the following for permission to reproduce material in this book:

© *The Guardian* for an extract from *What are fizzy drinks doing to our children?* by Sarah Boseley (9 January 2003) used on page 94; © Telegraph Group Limited for an extract from *Economic Law* by Patrick Minford (28 January 2002) used on page 66.

Every effort has been made to trace and acknowledge ownership of copyright. The publishers will be glad to make suitable arrangements with any copyright holders whom it has not been possible to contact.

Note about the Internet links in the book. The user should be aware that URLs or web addresses change regularly. Every effort has been made to ensure the accuracy of the URLs provided in this book on going to press. It is inevitable, however, that some will change. It is sometimes possible to find a relocated web page, by just typing in the address of the home page for a website in the URL window of your browser.

Orders: please contact Bookpoint Ltd, 130 Milton Park, Abingdon, Oxon OX14 4SB. Telephone: (44) 01235 827720. Fax: (44) 01235 400454. Lines are open from 9.00–6.00, Monday to Saturday, with a 24 hour message answering service. You can also order through our website www.hodderheadline.co.uk.

British Library Cataloguing in Publication Data
A catalogue record for this title is available from the British Library

ISBN (10) 0 340 846 37 2
ISBN (13) 978 0340 84637 7

First Published 2003
Impression number 10 9 8 7 6 5 4
Year 2009 2008 2007 2006

Cover photo the central business district of Singapore (by Michael Hill)
Produced by Gray Publishing, Tunbridge Wells, Kent
Printed in Great Britain for Hodder Education, a member of the Hodder Headline group, 338 Euston Road, London NW1 3BH by CPI Bath

Contents

1 The Development of the Global Economy

KEY WORDS

Internationalisation: the extension of economic activities across national boundaries. It is essentially a quantitative process, which leads to a more extensive geographical pattern of economic activity. The phase preceding globalisation.
Globalisation: the increasing interconnectedness of the world economically, culturally and politically. The current phase developing out of internationalisation.
Capitalism: the social and economic system, which relies on the market mechanism to distribute the factors of production (land, labour and capital) in the most efficient way.
Anti-capitalism: a broad term, which can cover any challenge to capitalism as the best or only way to organise the world. It was given media prominence during the 1999 WTO summit in Seattle where a wide range of organisations protested against the workings of the international economic system.
Debt: money owed by a country either to another country, private creditors (e.g. commercial banks) or international agencies such as the World Bank or IMF.
Structural adjustment: a set of policy changes countries are required to undertake in order to receive loans through the IMF and the World Bank.
Glocalisation: the notion of linking the global and local scales by thinking globally but acting locally.

1 Tasmanians know that they live on one planet because other people's aerosol sprays have caused a carcinogenic hole in the ozone layer over their heads, because their relatively high rate of unemployment is due to a slump in the international commodities markets, because their chil-
5 dren are exposed to such edifying role models as *Robocop* and *The Simpsons*, because their university is infested by the managerial cultures of strategic planning, staff appraisal and quality control, just like everyone else's, because British TV-star scientists may drop in for a week to save their environment for them, and because their gay community may
10 at long last be able to experience freedom of sexual expression because it has appealed to the human rights conventions of the United Nations.

*Malcolm Waters (*Globalization, *1995), commenting on global influences in Tasmania, which he sees as being at the spatial edge of human society.*

1 If anyone tells you geography doesn't matter anymore, just think how different your life would be had you been born in an African village. Or consider how Chinese immigrants in the USA lead different lives to

those of their comrades in China. Whereas a majority of Americans use
5 the Internet, half the world has never made a phone call.

*Philippe Legrain (*Open World*, 2002), stressing the continuing importance of geographical location.*

1 What is Globalisation?

While the word 'global' is over 400 years old, common usage of the word 'globalisation' did not commence until about 1960. In 1961 *Webster* became the first major dictionary to give a definition of globalisation. However, the word was not recognised as academically significant until the early- to mid-1980s. Since then its use has increased dramatically. The magazine *Newsweek* recently quoted a search of more than 40 major English language newspapers and magazines finding 158 stories that mentioned globalisation in 1991, and 17,638 in 2000. The famous American university Yale has established a Center for the Study of Globalization while some other universities have appointed professors of globalisation.

Some see the concept of globalisation as the key idea by which we understand the transition of human society into the third millennium. It is a truly international concept. In France the word is *mondialisation*, In Spain and Latin America it is *globalizacion* and in Germany *globalisierung*.

Malcolm Waters, the Australian sociologist, in his major contribution to the subject cites Roland Robertson as the key figure in the formalisation of the concept of globalisation. Robertson (1992) defined the term as follows:

> Globalisation as a concept refers both to the compression of the world and the intensification of consciousness of the world as a whole.

The idea of an intensification of global consciousness is a relatively new concept. In the process of arriving at his own definition Waters argues that the best approach might be to try to predict what a fully globalised world would look like. Its main characteristics would be a single global society and culture in which territoriality will disappear as an organising principle for social and cultural life. However, within this broad framework there will be a high degree of tolerance for diversity and individual choice. Waters sees flows of goods, people and ideas linking together previously homogeneous cultural niches forcing each to 'relativise'(compare, contrast and position) itself to others. Thus, it is a differentiating as well as a homogenising process, pluralising the world by recognising the value of cultural niches (different national cultures or minority cultures within nations). From this position Waters defines globalisation as:

> A social process in which the constraints of geography on economic, social and cultural arrangements recede, in which people become

> increasingly aware that they are receding and in which people act accordingly.

According to Waters, globalisation has become not only a major historical process that impacts on culture but the central substance of contemporary culture. It brings the centre to the periphery through the rapid and continuous transmission of Western culture but it also brings the periphery to the centre through the flow of economic migrants.

Globalisation is a process that people in many parts of the world are concerned about because it appears to justify the spread of Western culture and capitalism. However, as Robertson argues, it does not imply that every society must become Westernised and capitalist but rather that individual societies must establish (relativise) their positions in relation to the capitalist West.

Philippe Legrain, who is concerned that many of our worries about globalisation are unfounded, states:

> This ugly word is shorthand for how our lives are becoming increasingly intertwined with those of distant people and places around the world – economically, politically and culturally. These links are not always new, but they are more pervasive than ever before.

For others the nature of their job influences their thinking. For example for Kofi Annan, the Secretary-General of the United Nations, globalisation means 'world inclusivity' while Bill Gates, the President of Microsoft, views it as 'the world united by the Web'. There are, of course, many other views as the concept has found appeal across a wide range of intellectual interests.

Chinatown in San Francisco: a global city

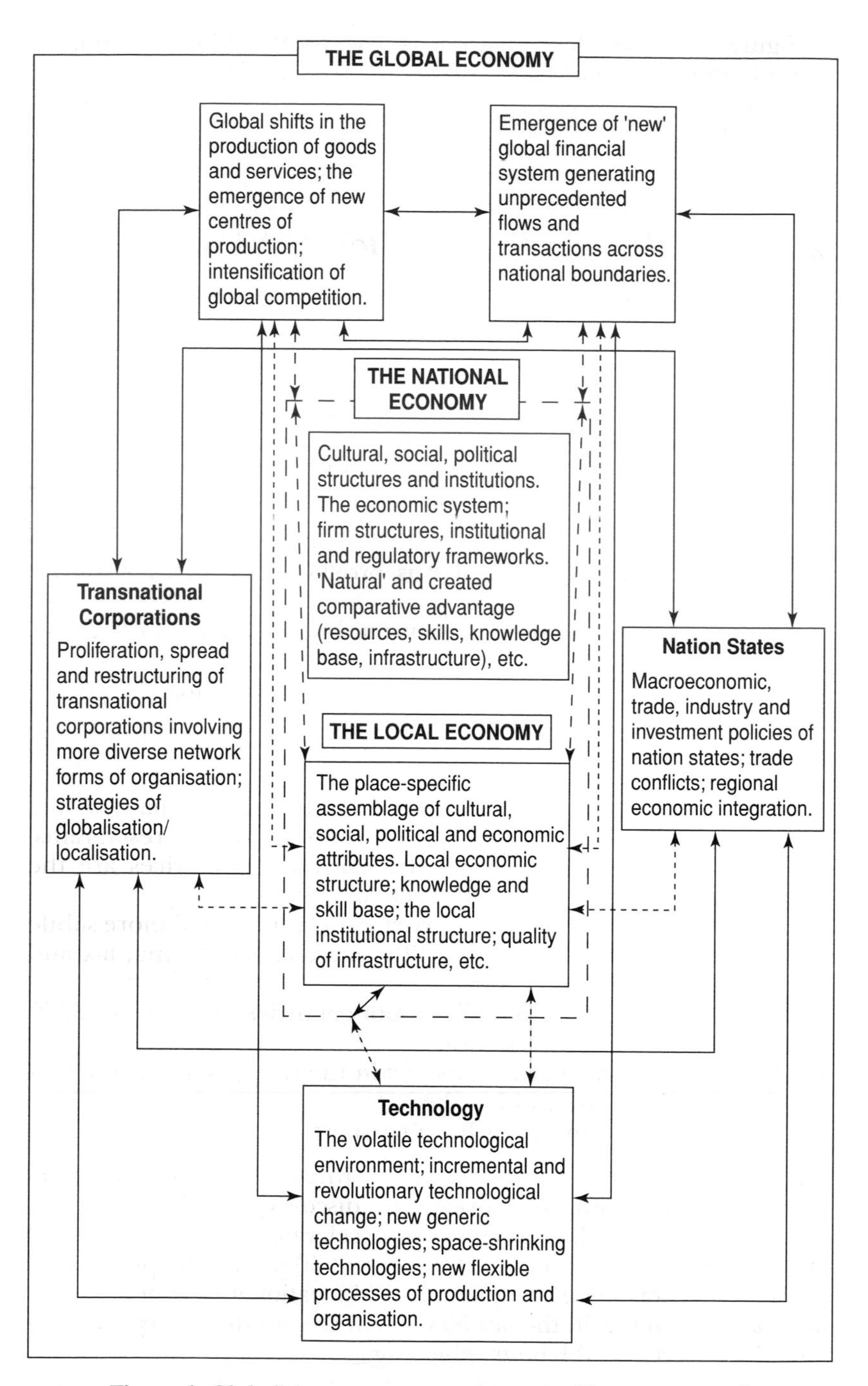

Figure 1 Globalising processes as a system of interconnected elements and scales. Source: *Global Shift* by P Dicken

Figure 1 shows Peter Dicken's view of the global economy. Transnational corporations and nation states are the two major shapers of the global economy. They are embedded within a triangular nexus of interactions consisting of firm–firm, state–state and firm–state relationships.

2 Internationalisation or Globalisation?

However, not everyone agrees that the process is occurring. Critics of the globalisation thesis argue that what has been happening in recent decades is not a new process but simply an extension of internationalisation. Alan Rugman, the Oxford management scholar, says that globalisation is an exaggeration of what has taken place in the international economy. Rugman emphasises the regional nature of so much international economic activity with a significant number of countries largely isolated from the so-called global economy. He and others point to facts such as:

- Over 70% of the world's GDP is produced domestically within national boundaries.
- Much of the change that has occurred is in increased trade shares among the rich nations.
- There has been very little increase in trade shares for the world's non-industrialised economies.
- Nearly 90% of what Americans consume is produced within the 50 US states.
- Most of what we consume cannot be traded. Since many of the services we use (e.g. hairdressers, shop assistants, restaurants, gardeners) have to be provided locally. These services are the fastest growing sector of MEDC economies.
- Even if all the remaining trade barriers were abolished more subtle barriers would remain such as differences in accounting, tax and regulatory standards.
- In 2000 imports from non-OPEC poor countries came to only 3.9% of GDP in rich OECD countries.
- Three-quarters of what Britons buy in the shops is made domestically and 90% is from within the EU.
- Only one in ten Americans has a passport.

The internationalisation of economic activities is not a new phenomenon. It began with the voyages of discovery of the European maritime powers. The process was accelerated by the spread of industrialisation. However, until the post-1950 period the production process itself was mainly organised within national economies. This has changed rapidly in the last 50 years or so with the emergence of a new global division of labour reflecting:

- A change in the geographical pattern of specialisation at the global

scale with the fragmentation of many production processes across national boundaries.
- The increasing complexity of international trade flows as this process has developed.
- The emergence of an increasing number of Newly Industrialised Countries (three generations can be recognised).
- The integration of the Soviet Union and its Eastern European satellites into the capitalist system.
- The opening up of other economies, particularly those of China and India.

It is the nature and scale of these changes that has justified the use of this new term 'globalisation' to distinguish the current phase that began in about 1970 from what was happening before (Figure 2). Globalisation is not an end product, at least not at the moment, but a process that has gathered momentum, albeit with interruptions. Peter Dicken in *Global Shift* states that 'both old and new industries are involved in this re-sorting of the global jigsaw puzzle in ways which also reflect the development of technologies of transport and communications, of corporate organisation and of the production process'.

Some writers argue that the level of integration before 1914 was similar to what it is today. However, Peter Dicken makes the distinction between the 'shallow integration' of the pre-1914 period and the 'deep integration' of the present period. Today linkages between national economies are increasingly influenced by cross-border value adding activities within TNCs and within networks established by TNCs. Globalisation processes are qualitatively different from internationalisation processes involving the functional integration of internationally dispersed activities. However, both processes co-exist.

	Phase 1		***Phase 2***	
	Fordism	Internationalisation	Post-Fordism	Globalisation
Production	Assembly line	MNCs	Flexible specialisation	Localisation/ regionalisation
Accumulation	Mass consumption	NIDL	Instantaneity/ niche marketing	Time–space compression
Regulation	Stoanism/ corporatism	Economic IGOs	Open factor markets	Electronic financial systems
Societalisation	Standardised nuclear family	Population & development programmes	Hyper-individuation	Economic human rights

Figure 2 Fordism and internationalisation, post-Fordism and globalisation. Source: *Globalization* by M Waters

'Globalisation has the potential to bring major improvements in productivity, innovation and creativity. But it's being overshadowed by a corporate-led plan for economic integration which threatens to undermine the whole project. Instead of helping build a better world for all, the current free-market model is eroding both democracy and equity'

W Ellwood (The No-Nonsense Guide to Globalisation)

'Globalisation is a myth due to economic illiteracy. It comes down to mass ignorance'.

Alan Rugman, Oxford University management scholar

'The world's corporate and political leadership is undertaking a restructuring of global politics and economics that may prove as historically significant as any event since the Industrial Revolution. This restructuring is happening at tremendous speed, with little public disclosure of its profound consequences affecting democracy, human welfare, local economics, and the natural world.'

The International Forum on Globalisation

'Although there are undoubtedly globalising forces at work we do not have a fully globalised world economy.'

Peter Dicken (Global Shift)

'Globalisation does not necessarily imply homogenisation or integration. It merely implies greater connectedness and de-territorialisation.'

Malcolm Waters (Globalization)

Figure 3 Some views on globalisation

According to Waters the terms 'post-Fordism' and 'economic globalisation' are alternative descriptors for a single, general set of processes of change in the economy.

Globalisation theory does not imply that the state is disappearing but that its sovereignty and its potency are being diluted. There can be little doubt that nation states will continue to be key players in the global economy. Also, although the impact of the transport and communications revolution cannot be underestimated, both geographical distance and place remain important components of decision-making.

3 The Dimensions of Globalisation

The ties that bind the world together are first the economic ones (Figure 4) of trade, investment and migration. The movement of goods, money and people around the globe brings far off places closer together. The World Trade Organisation (the successor of the GATT) has played a vital (and controversial) role in the increase in world trade. Relationships are also political. The increasing role played by the United Nations around the world epitomises the

Dimension	Characteristics
Economic	Under the auspices of first GATT and latterly the WTO, world trade has expanded rapidly. Transnational corporations have been the major force in the process of increasing economic interdependence, and the emergence of different generations of newly industrialised countries has been the main evidence of success in the global economy. However, the frequency of 'anti-capitalist' demonstrations in recent years shows that many people have grave concerns about the direction the global economy is taking. Many LEDCs and a significant number of regions within MEDCs feel excluded from the benefits of globalisation.
Urban	A hierarchy of global cities has emerged to act as the command centres of the global economy. New York, London and Tokyo are at the highest level of this hierarchy. Competition within and between the different levels of the global urban hierarchy is intensifying.
Social/cultural	Western culture has diffused to all parts of the world to a considerable degree through TV, cinema, the Internet, newspapers and magazines. The international interest in brand name clothes and shoes, fast food and branded soft drinks and beers, pop music, and major sports stars has never been greater. However, cultural transmission is not a one-way process. The popularity of Islam has increased considerably in many Western countries as has Asian, Latin American and African cuisine.
Linguistic	English has clearly emerged as the working language of the 'global village'. Of the 1.9 billion English speakers, some 1.5 billion people around the world speak English as a second language. In a number of countries there is great concern about the future of the native language.
Political	The power of nation states has been diminished in many parts of the world as more and more countries organise themselves into trade blocs. The European Union is the most advanced model for this process of integration taking on many of the powers that were once the sole preserve of its member nation states. The United Nations has intervened militarily in an increasing number of countries in recent time, leading some writers to talk about the gradual movement to 'world government'. On the other side of the coin is the growth of global terrorism.
Demographic	The movement of people across international borders and the desire to move across such borders has increased considerably in recent decades. More and more communities are becoming multicultural in nature.
Environmental	Increasingly, economic activity in one country has impacted on the environment in other nations. The long-range transportation of airborne pollutants is the most obvious evidence of this process. The global environmental conferences in Rio de Janeiro (1992) and Johannesburg (2002) is evidence that most countries see the scale of the problems as so large that only coordinated international action can bring realistic solutions.

Figure 4 The dimensions of globalisation

developing co-operation between nations even if the outcome of peacekeeping operations sometimes fails to match the original objective. The post-Second World War period has also witnessed the rapid development of trade blocs. Most countries in the world are now in some sort of trade agreement with their regional neighbours. The European Union is the most advanced model of this important phenomenon.

Another relatively recent phenomenon has been the growing importance of cross-border pressure groups such as Friends of the Earth and Amnesty International, elements of which has come to be known as 'global civil society'. The combined actions of these organisations and others have resulted in a growing framework of international rules on trade, the environment, human rights, war and other aspects of international relationships. However, the process is not always easy.

In America, conservatives see the creation of the first global criminal court, the International Criminal Court (ICC), as another step towards 'world government' that threatens US sovereignty. Thus, the USA was not among the 120 nations that initially endorsed the creation of the court at a 1998 conference in Rome. Although President Clinton eventually signed so that the USA could remain engaged in negotiations, once it became clear that the court would come into existence without the changes sought by Washington President Bush withdrew the USA from the Rome Treaty. Subsequently, Congress passed legislation that would authorise military action to free any American taken into custody on the direction of the ICC.

Globalisation not only pulls upwards in terms of nation states losing some of the power they once had as international linkages become more and more important, it also pushes downwards creating new pressures for local autonomy. According to the American sociologist Daniel Bell, nations are too small to solve the big problems but also too large to solve the small ones.

In a bid to offset criticism of globalisation, the world's first 'glocalisation' conference was held in Rome in May 2002, attended by mayors from a large number of the world's cities. The conference officially established the 'Glocal Forum', an organisation that will be at once global and local. Glocal Forum's president Uri Savir said in an opening address 'When we talk of globalisation, our thoughts go to the gap, the divide between the powerful and the weak. It is up to officials such as mayors to help make the world a more inclusive place from the local level – the city-upwards'.

The mixing of cultures is another important dimension of globalisation. This has occurred through:

- migration
- the rapid spread of news, ideas and fashions through the media, trade and travel, and the growth of global brands, such as Nike, Coca Cola and McDonald's, that serve as common reference points. The

terms 'Americanisation' and 'McDonaldisation' are often used to describe global consumer culture. These terms imply that the consumer culture that was developed in the USA in the middle of the 20th century has been spread by the mass media across the world.

In February 2003 George Tenet, the director of America's CIA was quoted as saying that globalisation had been 'a profoundly disruptive force for governments to manage'. He emphasises the stresses that Arab governments in particular were facing, particularly on the cultural front, without reaping the economic benefits. He also cited the process as a threat to US security.

4 The Development of Globalisation

When did globalisation begin? Three hypotheses have been proposed by various writers:

- that the process has operated since the dawn of history, increasing in its effects over time, with a sudden and recent acceleration
- that globalisation is cotemporal with modernisation and the development of capitalism, and that there has been a recent acceleration from the 1970s
- that it is a recent phenomenon associated with other social processes called post-industrialisation and post-modernisation.

The third of these hypotheses is the approach taken in this book. Globalisation developed out of internationalisation, which is the foundation of the second hypothesis. The argument here is that the change that has occurred since the 1970s is much more than an 'acceleration' of economic development.

Most theories of globalisation stress the economic foundations of the phenomenon. Globalisation is undoubtedly the direct consequence of the expansion of European culture across the planet through settlement, colonisation and cultural replication since the 15th century. The voyages of Cristobal Colon (Christopher Columbus) and others opened the door to over 450 years of European colonialism. The main impetus for this expansion was economic gain. The forerunners of today's transnational corporations (TNCs), the chartered trading companies that emerged in Europe from the 15th century onwards, were to play a crucial part in this process. For example, in 1628 King Charles I of England chartered the Massachusetts Bay Company in order to colonise the New World. When Americans rebelled against Britain it was partly out of fear of the power of crown trading companies. These trading companies operated in other parts of the world too. The economic historian Harold James stated that 'The British East India Company was the McDonald's of the day'. However, it must be remembered that the trading companies of this era were primarily involved in trade and exchange rather than production. The first companies to engage in production outside of their domestic base did not emerge until the second half of the 19th century.

Karl Marx saw the bourgeoisie spreading its tentacles around the world because of the need of a constantly expanding market for its products. In the process the bourgeoisie was recreating the rest of the world in its own image.

Although the global spread of capitalism began in the 16th century its pace has accelerated through time and it is currently in the most rapid phase of development. According to Wallerstein (1990):

> The capitalist world-economy has seen the need to expand the geographic boundaries of the system as a whole, creating thereby new loci of production to participate in its axial division of labour. Over 400 years, these successive expansions have transformed the capitalist world economy from a system located primarily in Europe to one that covers the entire globe.

The process has been driven partly by cheaper, easier and faster transportation. In 1850 it took nearly a year to sail or send a message around the world. Now an email can be sent anywhere almost instantaneously. However, globalisation involves more than technological change as it is also a political choice, consciously opening national borders to foreign influences. Philippe Legrain states 'If governments wanted to, they could put globalisation into reverse again'. However, other writers would argue that this is easier said than done.

A key period in the process of internationalisation occurred between 1870 and 1914 when:

- transport and communications networks expanded rapidly around the world
- world trade grew significantly with a considerable increase in the level of interdependence between rich and poor nations
- there were very large flows of capital from European companies to other parts of the world.

Between 1800 and 1913 international trade grew from 3% to 33% as a proportion of world product. It tripled between 1870 and 1913. At this time the world trading system was dominated and organised by four nations: Britain, France, Germany and the USA. Such was the extent of global interaction at this time that capital transfers from North to South were actually greater at the end of the 1890s than at the end of the 1990s. By 1913 exports accounted for a larger share of global production than they did in 1999. 'What an extraordinary episode in the economic progress of man was that age which came to an end in August 1914', wrote John Maynard Keynes.

As industrialisation spread across the globe it carried modernisation with it, reducing the differences between societies. As societies progressively sought the most effective technology of production their social systems progressively adapted to that technology. Industrial societies are organised spatially into cities. The process of urbanisation that commenced in the developed world in the late 18th century has been replicated in the developing world since the 1950s.

The global shocks of the First World War and the Great Depression put a stop to the first great wave of internationalisation that had begun in the 19th century. Countries increasingly turned inwards. For example, international lending fell by over 90% between 1927 and 1933. It was not until the 1950s that international interdependence was back on track.

Since the 1950s, world trade has grown consistently faster than world GDP, although even by 1990 the level was unremarkable compared with the late 19th and early 20th centuries. However, today's globalisation is very different from the global relationships of 50 or 100 years ago. The character of the global economy and its impact on people and the natural world is totally different today.

TNCs are the key players in the economics of globalisation. The social side of the process has been driven by the culture of consumerism. In the 19th century it was Britain that took the lead in promoting capitalism as the global system; in the 20th century it has been the USA. The growth of global companies has been particularly rapid in recent decades. In 1975 there were still only 7000 transnational corporations compared with more than 60,000 today.

The dimension of the world economy that is most globalised is the market for raising loans and capital. A number of stages in this process can be recognised (Figure 5). Gilpin (1987) ended his third stage in 1985, his time of writing, but clearly the third stage is still continuing. The new international financial system is characterised by rapid 24-hour global transactions.

In contrast to financial markets, labour markets are far behind in the process of globalisation largely because of the restrictions governments impose on immigration. However, the natural attachment that people feel for their 'home' region is also a significant factor in the immobility of labour. In a genuinely globalised market there would be no restrictions on the movement of labour. Nevertheless, labour movement across international borders is occurring. A significant feature of the current situation is the way peoples from the rich and poor world are beginning to mingle in global cities. According to Waters 'Under globalisation, migration has brought the third world back to the global cities, where its exploitation becomes ever more apparent'.

The origin and continuing basis of global interdependence is trade. Frequently established on a colonial basis, the economic and cultural ties still remain in the post-colonial era. By 1975 a strongly internationalised economy had been established. In the previous 25 years world trade had increased tenfold. World production was dominated by about 50 TNCs mainly based in the USA, Europe and Japan.

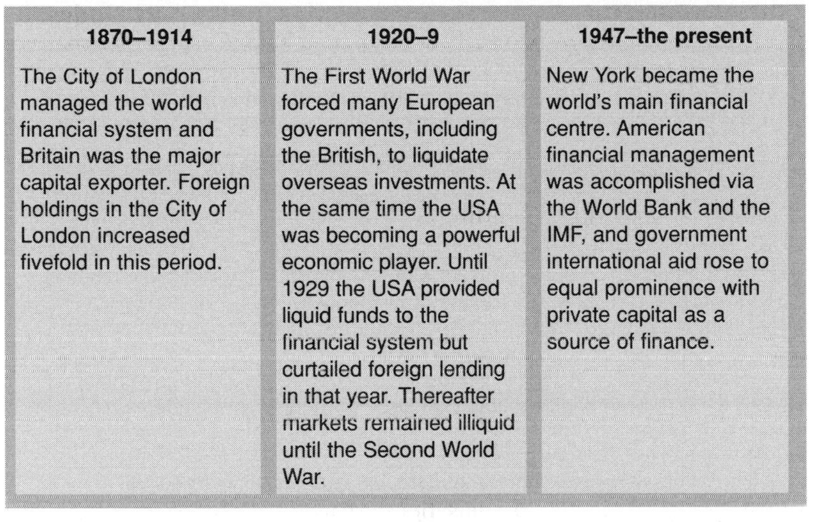

1870–1914	1920–9	1947–the present
The City of London managed the world financial system and Britain was the major capital exporter. Foreign holdings in the City of London increased fivefold in this period.	The First World War forced many European governments, including the British, to liquidate overseas investments. At the same time the USA was becoming a powerful economic player. Until 1929 the USA provided liquid funds to the financial system but curtailed foreign lending in that year. Thereafter markets remained illiquid until the Second World War.	New York became the world's main financial centre. American financial management was accomplished via the World Bank and the IMF, and government international aid rose to equal prominence with private capital as a source of finance.

Figure 5 Gilpin's stages in the development of international financial markets

World trade now accounts for 25% of GDP, double its share in 1970. However, trade figures understate the degree of international interdependence. Foreign direct investment, cross-border mergers and financial markets have an even bigger impact.

Key developments in recent decades have been:

- The emergence of fundamentalist free-market governments in the USA (Ronald Reagan) and Britain (Margaret Thatcher) around 1980.
- The disintegration of the state-run command economies in the former Soviet Union and its East European satellites. No significant group of countries any longer stands outside the free market global system. As Anthony Giddens states 'that collapse wasn't just something that happened to occur. Globalisation explains both why and how Soviet communism met its end'.
- The deregulation of world financial markets.

Until the early 1970s the Soviet Union and Eastern Europe were roughly comparable to the West in terms of growth rates. After this time they fell rapidly behind. Soviet communism, with its emphasis upon state-run enterprise and heavy industry, could not compete in the global electronic economy. In addition, the ideological and cultural control upon which communist political authority was based could not survive in an era of global media.

Prior to the deregulation of global financial markets the activities of banks, insurance companies and investment dealers had been confined largely within national boundaries. With the removal of

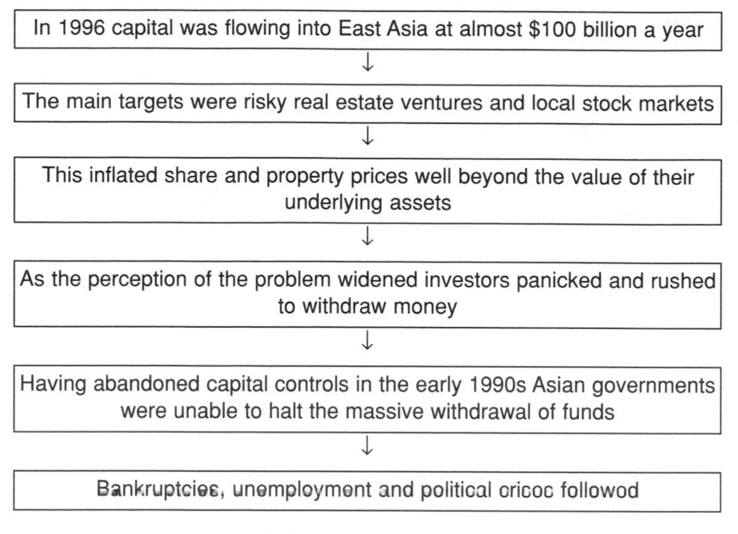

Figure 6 East Asian financial crisis

regulations the financial sector scanned the globe for the best returns. According to W Ellwood 'In this new relaxed atmosphere finance capital became a profoundly destabilising influence on the global economy with an increasing level of speculation as opposed to 'firm' investment'. Recent UN studies show a direct correlation between the frequency of financial crises and the increase in international capital flows during the 1990s. Nervous short-term investors who withdraw capital at the first hint of a problem can cause a vicious downward spiral to a country's economy, which can be very difficult for a government to counter (Figure 6). Critics of globalisation often use the term 'global casino' to refer to rapid movement of speculative money around the world.

The impact of the economic crisis that hit East Asia in the late 1990s was severe. More than 400 Malaysian companies declared bankruptcy between July 1997 and March 1998, and in Indonesia 20% of the population were pushed into poverty through unemployment. The East Asian crisis was a major blow to the promise of economic globalisation. It was now apparent that the global economy was more fragile than most had imagined. This crisis, added to the other perceived problems of globalisation, led to the mass public protests in Seattle (1999) and Prague (2000).

Some economists have voiced concerns that increasing globalisation could lead to bigger economic booms and busts. Figure 7(a) shows that the dispersion of growth rates across 41 economies fell to its lowest level in at least 30 years. It seems that different economies'

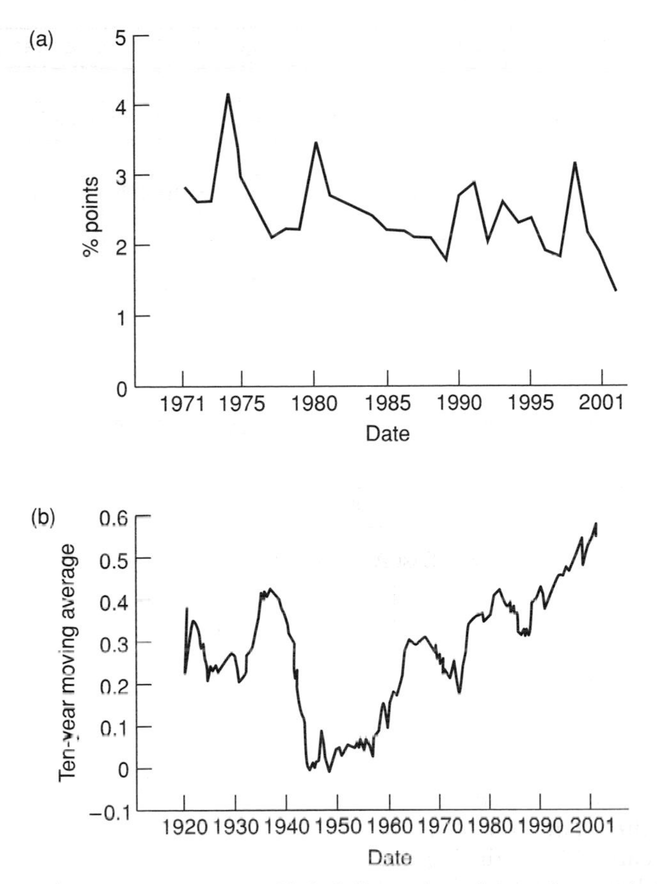

Figure 7 (a) Convergence. Global dispersion of growth rates (standard deviation of GDP growth rates in 41 economies, weighted by GDP). Source: JP Morgan. From *The Economist*, 28 September 2002.
(b) Dancing to Wall Street's tune. Average correlation of US and European (Germany, France and Britain) stockmarkets, ten-year moving average. Sources: Goetemann, Li and Rowenhurst, NBER Working Paper no. 8612. From *The Economist*, 28 September 2002

business cycles are becoming more closely correlated over time, possibly because of greater economic integration. This may well make it more likely that a crisis in one or two major economies will have more of a knock-on effect on other economies than in the past. The increasing correlation between major economies is also evident in other ways. Cross-border trading in shares has increased considerably over the past decade (Figure 7b), creating a global equity market. The Yale economist William Goetzmann calculates that in recent years the correlation between the markets of the USA, Germany, Britain and France is closer than at any time in the past century.

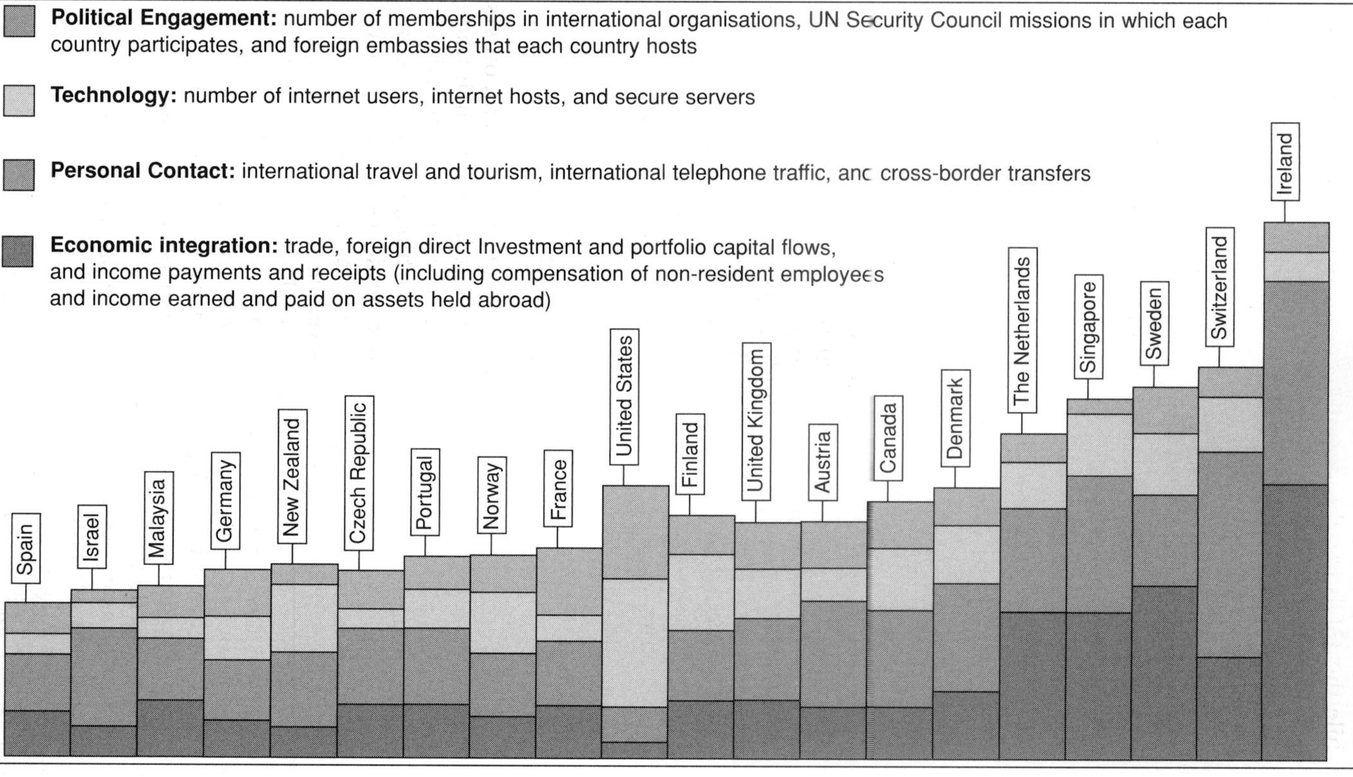

Figure 8 The most globalised countries in the world

While the 1980s was the Japanese decade according to most indicators, the 1990s turned out to be a decade of unexpected prosperity for the USA – what some of the American journals have called 'the New Economy'. The assertion is that the New Economy is the beginning of a major new wave of innovation. Historically, periods of major innovation have brought substantial increases in living standards. The view is that the latest wave could make it much easier to address some of the vexing social and environmental problems that affect individual countries and global society in general.

Which are the most globalised countries? A number of attempts have been made by various journals and organisations to compare nations in terms of their global connections, of which Figure 8 is one version. Here four components of globalisation are analysed.

5 Which are the Major Global Economies?

Figure 9(a) shows the relative size of the top ten global economies according to the traditional measure, gross domestic product (GDP). However, more and more organisations such as the UN and the International Monetary Fund are publishing GDP data at purchasing power parity (PPP). Once differences in the local purchasing power of currencies are taken into account, China's economy is just over half the size of the USA's (Figure 9b). The other major emerging economy, whose relative importance increases significantly once output is measured on a PPP basis, is India.

6 How Inclusive is Globalisation?

Global integration is spatially selective: some countries benefit, others it seems do not. A few developing countries have increased their trade substantially. These countries have attracted the bulk of foreign direct investment. Such low-income 'globalisers' as China, Brazil, India and Mexico have increased considerably their trade-to-GDP ratios. GDP per capita in these economies grew by an average of 5% a year during the 1990s, compared with 2% in the developed countries. However, on the other side of the coin are the two billion people who live in countries that have become less rather than more globalised (in an economic sense) as trade has fallen in relation to national income. This group includes most African and many Muslim countries. In these 'non-globalising' countries income per person fell by an average of 1% a year during the 1990s.

The conditions of the world's poorest people are a serious cause for concern. For example:

- One in five of the world's population live on less than a dollar a day, almost half on less than two dollars a day.
- More than 850 million people in poor countries cannot read or write.

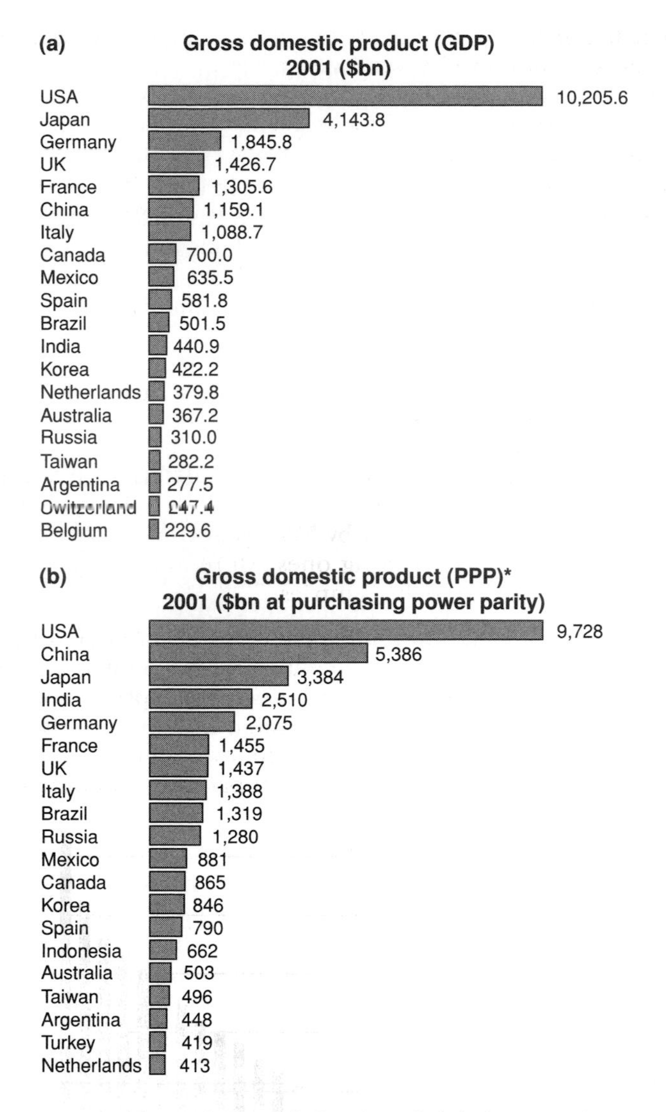

Figure 9 The relative size of global economies

- Nearly a billion people do not have access to clean water and 2.4 billion to basic sanitation.
- 11 million children under five die from preventable diseases each year.
- In the 1990s the share of the poorest fifth of the world's population in global income dropped from 2.3% to 1.4%. On the other hand the proportion taken by the richest fifth rose from 70% to 85%.

- In Sub-Saharan Africa, 20 countries have lower incomes per head in real terms than they did two decades ago.
- At the beginning of the 19th century, the ratio of real incomes per head between the world's richest and poorest countries was three to one. By 1900, it was ten to one. By 2000, it had risen to 60 to one. Today, world GDP per capita (purchasing power parity) is around $6000, with the richest nation at $29,000 and the poorest at $500.

Philippe Legrain sees globalisation as '*the only realistic route out of poverty for the world's poor*'. He stresses that the poor are generally getting richer in globalising countries. A number of academic studies seem to support his view:

- J Frankel and D Romer writing in the *American Economic Review* (June 1998) found that a 1% rise in the share of imports and exports in national income raised GDP by 2% or more.
- World Bank studies by D Dollar and A Kraay have reached a similar conclusion. They found that whereas in LEDCs that are globalising GDP per person increased by 5% a year in the 1990s, it rose by 1.4% a year in non-globalising ones. Moreover, the globalisers are catching up with the rich countries.

However, others from a variety of different disciplines blame the rules of the global economic system for excluding many countries from its potential benefits. Many single out debt as the major problem for the world's poorer nations. By 1999 the foreign debts of LEDCs had reached nearly $3000 billion (Figure 10). An ever increasing proportion of new debt has been to service interest payments on old debts.

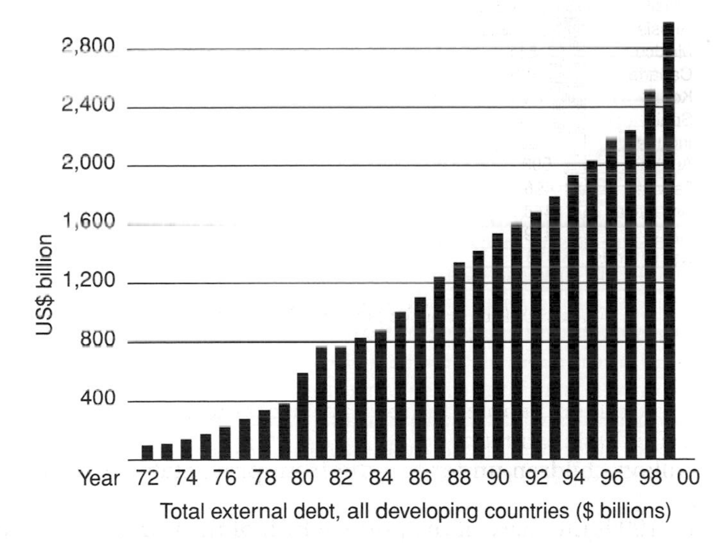

Figure 10 Total external debt, all developing countries ($ billions). Source: World Development Report 1999–2000, World Bank

While Legrain argues that economic growth through trade is the only answer, critics say that MEDCs should do more to help the poor countries through debt relief and by opening their markets to exports from LEDCs.

Although many people are concerned that Western influences are dominating the diverse cultures of the world, examples of 'reverse colonisation' are becoming more and more common. Reverse colonisation means that non-Western countries influence developments in the West. Examples include:

- the sale of Brazilian TV programmes to Portugal, Italy and other MEDCs
- the location of South Korean and Taiwanese factories in Britain
- the emergence of a globally-oriented high-tech sector in India.

7 The International Monetary Fund, the World Bank and GATT/WTO

The International Monetary Fund and the World Bank (both are United Nations agencies) were brought into being at a conference designed to plan a new economic structure for the post-war period, held at Bretton Woods, New Hampshire in 1944. The initial role of the World Bank was to assist in the funding of reconstruction in the countries decimated by war, while the IMF would ensure that the process would take place in a stable economic climate. A country running short of foreign currency reserves that it needed to maintain its currency exchange rate could turn to the IMF for help. IMF funds come from the contributions or 'quotas' of its member countries. Voting power on the IMF is in proportion to the size of a country's quota, with the USA holding 17% of the total votes. As any major change to IMF policy requires 85% backing the USA is able to block by itself any proposed change it might not like.

Countries usually apply for funding from the IMF when they are unable to obtain funding from other sources. IMF money is designed to prevent the disruption to the international financial system that would occur through a country failing to meet its commitments to other nations. Along with funding the IMF is also able to renegotiate the terms of debt on behalf of nations in financial difficulties. To prevent the situation reoccurring the IMF will usually impose conditions, in the form of a 'stabilisation programme', on its financial assistance. The objective is 'structural adjustment', changing the fundamental conditions of the economy to make it more competitive and less likely to return to crisis. It is the nature of these conditions (Figure 11) that has caused so much controversy about the way in which the IMF operates. Opposition groups often find other meanings for the initials IMF. One that has been doing the rounds for a while is the International Meddlers Federation.

The IMF Stabilisation Programme

The conditions of the IMF stabilisation programme are often criticised as being anti-developmental. They administer a sharp shock to the domestic economy that its peoples can ill-afford. The tightening of domestic economic policy is likely to send the economy into recession by lowering domestic demand. Firms which cannot find new export markets for their products will therefore experience difficulties and will also face higher costs for imported raw materials (due to the exchange rate devaluation). Workers are likely to lose their jobs as a result and a downward multiplier effect results. Evidence suggests that the adjustments insisted on by the IMF hit poor and middle income families hardest. One example of this effect in action is the cuts in government spending that form part of the fiscal tightening. Government spending on projects to meet basic needs is curtailed. The IMF is essentially a free-market institution and the imposition of such stringent conditions reflects a belief that the economy can quickly begin to grow once the economic fundamentals have been put in place. However, even the IMF's own survey of past stabilisation programmes acknowledged in 1995 that 'few if any countries have shifted to a distinctly more rapid pace of growth backed by higher savings ratios.' The programmes have succeeded to some degree in over-coming short term balance of payments crises and hence stabilising the world economy. It is also clear that the IMF has fostered greater integration of the global economy, through its insistence on measures such as the reduction of protectionist barriers. Greater integration is undoubtedly a goal of the IMF, but it is not necessarily true that greater integration encourages development. The International dependence school of thought would view the IMF as an agent of developed countries, whose short term economic interests are best served by maintaining impoverishment in less developed countries.

Figure 11 The IMF Stabilisation Programme. Source: P Cramp, *Economic Development*, Anforme, 2001 (2nd Edition)

The International Bank for Reconstruction and Development, commonly known as the World Bank, borrows between $20 billion and $30 billion a year in a variety of currencies. This money has provided financing for more than 4000 development projects in 130 countries, through $300 billion in lending. When the reconstruction of Europe was complete the World Bank increasingly turned its attention to developing countries.

While the IMF focuses primarily on the international financial transactions of a country, the World Bank deals mainly with internal investment projects. For most recipient countries lending is at market rates of interest. However, in 1960 a branch of the World Bank known as the International Development Association (IDA) was formed. The IDA lends only to nations with a very low per capita income. For such countries loans are interest free and allow long repayment periods.

Four general phases in World Bank policy can be recognised:

- In the 1950s and 1960s World Bank projects in developing countries were mainly large scale (e.g. financing dams, airports, roads). Unfortunately, such schemes had only a limited impact on per capita incomes in the countries concerned.
- In the 1970s there was a shift in emphasis to smaller projects. The objective was to target projects that would act directly to alleviate poverty. The role of the IDA was crucial in this respect.
- In the 1980s the emphasis was on loans for structural adjustment and the promotion of the market economy. This remains an important aspect of the work of the World Bank.
- In the 1990s the World Bank began to promote sustainable development with most funding going to small-scale projects.

The World Bank has many critics. The US-based Heritage Foundation examined economic growth rates in the 85 countries that received World Bank International Development Association (IDA) loans in the 1980s and 1990s, and found that:

- Rather than helping the recipient countries, the loans pushed many into further debt, with new loans often being used to pay off old ones, the classic vicious circle.
- Recipient countries were more likely to experience a drop in per capita wealth than to achieve significant economic growth.

In spite of adverse publicity such as this the AAA-rated (the highest credit rating available) World Bank is highly sought after by global investors who buy bonds and in doing so provide the funds that the Bank distributes to development projects. However, the World Bank Bonds Boycott campaign is trying to deter investors from continuing their support of the World Bank, arguing that the conditions attached to World Bank loans have:

- crippled economic growth in recipient countries
- hindered development
- promoted dependency
- and increased poverty.

Critics argue that the rich nations use the World Bank to run other countries for the benefit of their merchant banks. Many countries and organisations are calling for the reform of the World Bank and the IMF. Others go further and argue for the abolition of these agencies and a complete restructuring of the world financial system.

The view that the protectionist policies of the 1930s should not be allowed to occur led to the establishment of another important international institution in 1947, the General Agreement on Tariffs and Trade (GATT). The objective of GATT was to gradually lower the barriers to trade, with free trade as its conceptual objective. GATT will be discussed in more detail in the next chapter.

8 The Rise of Global Civil Society

The term '*global civil society*' has become part of the literature concerning the process of globalisation. Anheier, Glasius and Kaldor identify four interpretations of this term:

- Protest groups who act as a counterweight to capitalism. Their aim is to 'civilise' globalisation.
- The human infrastructure that is needed for the spread of democracy and development.
- The efforts of groups like Save the Children and Médecins Sans Frontières to provide humanitarian assistance, and other signs of global solidarity with the poor or oppressed.
- The growing connectedness of citizens around the world.

The concept posits the existence of a social sphere above and beyond national, regional or local societies. Figure 12 summarises the main views on globalisation as seen by Anheier, Glasius and Kaldor.

Critics of the way that globalisation is operating highlight the following:

- the widening gap between rich and poor
- decision-making power being concentrated in fewer and fewer hands
- the erosion and loss of local cultures
- the destruction of biological diversity
- the increase in environmental problems
- the increase in regional tensions.

A major issue is the apparent control that a relatively small number of countries have over the major international organisations (Figure 13). Those who have major concerns about globalisation, the so-called anti-globalisation movement, come from a wide variety of backgrounds:

- Popular fears about the power of big business.
- Trade unionists worried about jobs filtering down to lower wage economics.
- Environmentalists who say that TNCs are disregarding the environment in the rush for profits and market share.
- Those fearful of the erosion of national sovereignty and culture.
- Small businesses afraid that they will become the victims of global economies of scale.
- Poverty campaigners who say that the West's gain has been at the expense of developing countries.

All these groups and others came together to demonstrate against the WTO in Seattle in December 1999 and at subsequent international conferences. Mark Ritchie, President of the Institute of Agriculture and Trade Policy described the demonstrations at the Seattle Conference as 'the first post-modern global gathering ... the nations

	Types of actors	Position on globlisation	Position on plant biotechnology	Position on global finance	Position on humanitarian intervention
Supporters	Transnational business and their allies	Favour global capitalism and the spread of a global rule of law	Favour plant biotechnology developed by corporations, no restrictions necessary	Favour de-regulation, free trade and free capital flows	Favour just wars for human rights
Rejectionists	Anti-capitalist social movements; authoritarian states; nationalist and fundamentalist movements	Left oppose global capitalism; right and left want to preserve national sovereignty	Believe plant biotechnology is 'wrong' and dangerous and should be abolished	Favour national protection of markets and control of capital flows Radical rejectionists want overthrow of capitalism	Oppose all forms of armed intervention in other states, intervention is imperialism or not our business
Reformists	Most INGOs; many in international institutions; many social movements and networks	Aim to civilise globalisation	Do not oppose technology as such but call for labelling information and public participation in risk assessment; sharing of benefits	Want more social justice and stability. Favour reform of international economic institutions as well as specific proposals like debt relief or Tobin tax	Favour civil society intervention and international policing to enforce human rights
Alternatives	Grass roots groups, social movements and submerged networks	Want to opt out of globalisation	Want to live own lifestyle rejecting conventional agriculture and seeking isolation from GM food crops	Pursue an anti-corporate life-style, facilitate colourful protest, try to establish local alternative economies	Favour civil society intervention in conflicts but oppose use of military force

Figure 12 Global civil society positions on globalisation. Source: *Introducing Global Civil Society*, Helmut Anheier, Marlies Glasius and Mary Kaldor

The United Nations:	*The International Monetary Fund and the World Bank*:	*The World Trade Organisation*:
the five permanent members of the UN security council each have the power of veto not only over decisions concerning war and peace, but also over all attempts to amend or review the UN charter.	altering the constitution of either body requires an 85% vote. The USA alone possesses 17% of the votes in each organisation.	in principle, every nation has an equal vote within the WTO. In practice the rich world shuts the poor world out of key negotiations.

Figure 13 Control of major international organisations

Global civil society: anti-war demonstration in London, 2003

of the South combined with representatives of civil society to write a new chapter in global governance'.

9 Globalisation and the Environment

Environmentalists argue that the economics of globalisation is concerned primarily with internal costs, largely ignoring external costs such as environmental impact. While companies make profits, society has to pay the bill. There is a great deal of evidence that the planet's ecological health is in trouble. Between 1950 and 2000 humankind consumed more of the world's natural capital than during the entire previous history of man. According to the ecologist Robert Ayres 'We may well be on the way to our own extinction'. Cuts enforced by the IMF have reduced spending on the environment in a number of countries.

In many countries there is a shortage of good agricultural land and increasing demands placed on limited water supplies. In some nations forests are disappearing at a rapid rate in an attempt to extend agricultural land. But, far too often, forest soils quickly degrade with such a change in use.

10 Has Globalisation Run Out of Steam?

Most people feel that it is inevitable that globalisation will continue to develop, but this cannot be taken for granted. As *The Economist* pointed out in a recent special report on the subject (2 February 2002)

> The lesson of the early 20th century, easily forgotten during the boom years of the 1990s, is that globalisation is reversible. It was derailed by war (in 1914) and by economic policy during recession (in the early 1930s). This time global integration might stall if the risk and cost of doing business abroad rises (perhaps as a consequence of heightened fears about security), or if governments once more turn their backs on open trade and capital flows. Either of these threats could prove decisive. The question is, will they?

A recent article in *Newsweek* entitled 'the New Buzzword: Globalony' argues that the extent of globalisation was over-exaggerated in the boom years of the 1990s and that the much more difficult economic and political climate of the first decade of the 21st century requires a more considered assessment of the interconnectedness of nations around the world. It seems that the term 'globalony' was coined in 1943 by US Congresswoman Clare Boothe Luce to criticise what Vice President Henry Wallace liked to call his 'global thinking', particularly a plan to promote world peace by building airports all over the world. The evidence of at least a temporary blip in the globalisation process is:

- World trade fell by 4% in 2001 after growing at an annual rate of 5% for a decade.
- In 2002 the USA took major protectionist steps backwards on steel and farm subsidies raising concern about the outcome of new free trade talks.
- Businesses have significantly reduced investment into new ventures abroad.
- Investors have been fleeing LEDC stocks since the financial crises in Asia and Russia in the late 1990s
- The growing concern that the international authorities (World Bank, IMF, US Federal Reserve), who seemed so omnipotent in the boom years, are unable to thwart an emerging crisis.

Summary

- Globalisation refers to the increasing interconnectedness of nations. There are many dimensions of the phenomenon including economic, cultural and political. Globalisation developed out of internationalisation.
- Internationalisation began with the voyages of European explorers around the world from the late 15th century.
- Key periods in the process of global integration occurred between 1870 and 1914, and from about 1960 to the present time.
- A major impetus to globalisation has been the deregulation of world financial markets. Critics see the rapid movement of speculative money around the world as a destabilising influence on the global economy.

- The International Monetary Fund and the World Bank play major roles in running the global economy. Many countries and organisations are critical about the way these two powerful bodies operate.
- The total external debt of the developing countries has risen significantly over the last three decades.
- The term Global Civil Society is increasingly used to sum up the many facets of international opposition to the way in which the global economy is run.
- There are growing concerns about the impact of globalisation on the environment.
- Some commentators argue that the extent of globalisation has been over-exaggerated.

Questions

1. **(a)** Define 'globalisation'.
   (b) Discuss the differences between internationalisation and globalisation.
   (c) Why do some academics argue that globalisation does not really exist?
2. Study Figures 4, 5 and 6.
   (a) Briefly examine the different dimensions of globalisation.
   (b) Outline the stages in the development of international financial markets.
   (c) Explain the causes and consequences of the East Asian financial crisis.
3. Study Figures 7 and 8.
   (a) Describe the trends shown in Figures 7(a) and (b).
   (b) Suggest why these trends have occurred.
   (c) What are the possible negative consequences of these trends?
   (d) Discuss the factors that have been used to rank the most globalised countries in the world in Figure 8.
   (e) Comment on the size and geographical distribution of these countries.
4. Study Figure 9.
   (a) Define: (1) Gross Domestic Product; and (2) Purchasing Power Parity.
   (b) Identify the differences between the two rankings.
   (c) Explain the differences you have identified in (b).
5. **(a)** What are the functions of the World Bank and the International Monetary Fund?
   (b) How has the policy of the World Bank changed since the 1950s?
   (c) Why are many commentators critical of the way the World Bank operates?
   (d) Outline the criticisms of the IMF stabilisation programme.

2 World Trade

KEY WORDS

General Agreement on Tariffs and Trade (GATT): an international agreement to reduce barriers to trade such as tariffs, quotas and subsidies. It was established in 1947 and succeeded by the World Trade Organisation in 1995.
World Trade Organisation: established in 1995 as a permanent organisation with far greater powers to arbitrate trade disputes than its predecessor (GATT).
Free trade: a hypothetical situation whereby producers have free and unhindered access to markets everywhere. While trade is more free today than in the past, governments still impose significant barriers to trade and often subsidise their own industries in order to give them a competitive advantage.
Trade bloc: a group of countries that share trade agreements between each other.
Protectionism: the institution of policies (tariffs, quotas, regulations) that protect a country's industries against competition from cheap imports.
The terms of trade: the price of a country's exports relative to the price of its imports, and the changes that take place over time.
Fair trade: a movement that that aims to create direct long-term trading links with producers in developing countries, and to ensure that they receive a guaranteed price for their products, on favourable financial terms.

World trade rules are a key part of the poverty problem; fundamental reforms are needed to make them part of the solution

Oxfam report published in May 2002.

The 1980s and 1990s saw the globalisation of the idea of the trading bloc as a means to the expansion of national trade.

Malcolm Waters

1 The Distribution of World Trade

Figure 14 shows the spatial distribution of world trade in merchandise and commercial services. In terms of individual countries, the USA is by far the largest exporter of merchandise. However, it dominates imports by a much greater margin, taking over one-fifth of the world total. This trade deficit is something that worries many economists as it is a situation that cannot be maintained indefinitely.

World merchandise exports by region, 2001 (billion dollars and percentage)

	Value	Share	
	2001	1990	2001
World	5984	100.0	100.0
North America	991	15.4	16.6
United States	731	11.6	12.2
Latin America	347	4.3	5.8
Mexico	159	1.2	2.6
Western Europe	2485	48.2	41.5
European Union (15)	2291	44.4	38.3
C./E. Europe/Baltic States/CIS	286	3.1	4.8
Central and Eastern Europe	129	1.4	2.2
Russian Fed.	103		1.7
Africa	141	3.1	2.4
South Africa	29	0.7	0.5
Middle East	237	4.1	4.0
Asia	1497	21.8	25.0
Japan	403	8.5	6.7
China	266	1.8	4.4
Six East Asian traders	568	7.8	9.5
Memorandum item:			
MAFTA (3)	1149	16.6	19.2
MERCOSUR (4)	88	1.4	1.5
ASEAN (10)	385	4.2	6.4

World merchandise imports by region, 2001 (billion dollars and percentage)

	Value	Share	
	2001	1990	2001
World	6270	100.0	100.0
North America	1408	18.3	22.5
United States	1180	14.8	18.8
Latin America	380	3.7	6.1
Mexico	176	1.2	2.8
Western Europe	2524	48.6	40.3
European Union (15)	2334	44.6	37.2
C./E. Europe/Baltic States/CIS	267	3.3	4.3
Central and Eastern Europe	159	1.4	2.5
Russian Fed.	54		0.9
Africa	136	2.8	2.2
South Africa	28	0.5	0.5
Middle East	180	3.0	2.9
Asia	1375	20.3	21.9
Japan	349	6.7	5.6
China	244	1.5	3.9
Six East Asian traders	532	8.0	8.5
Memorandum item:			
MAFTA (3)[a]	1578	19.3	25.2
MERCOSUR (4)	84	0.8	1.3
ASEAN (10)	336	4.6	6.4

[a] Imports of Canada and Mexico (1990–9) are valued f.o.b.

World exports of commercial services by region, 2001 (billion dollars and percentage)

	Value	Share	
	2001	1990	2001
World	1460	100.0	100.0
North America	299	19.3	20.5
United States	263	17.0	18.1
Latin America	58	3.8	4.0
Mexico	159	1.2	2.6
Brazil	9	0.5	0.6
Western Europe	679	53.1	46.5
European Union (15)	612	47.2	41.9
United Kingdom	108	6.9	7.4
France	80	8.5	5.5
Germany	80	6.6	5.5
Italy	57	6.2	3.9
Central and Eastern Europe, the Baltic	56	2.6	3.8
Africa	31	2.4	2.1
Egypt	9	0.6	0.6
South Africa	5	0.4	0.3
Middle East	33	2.0	2.2
Israel	11	0.6	0.8
Asia	303	16.8	20.8
Japan	64	5.3	4.4
Hong Kong, China	42	2.3	2.9
China	33	0.7	2.3
Korea, Rep. of	30	1.2	2.0
Singapore	26	1.6	1.8
India	20	0.6	1.4

World imports of commercial services by region, 2001 (billion dollars and percentage)

	Value	Share	
	2001	1990	2001
World	1445	100.0	100.0
North America	229	15.4	15.9
United States	188	12.0	13.0
Latin America	71	4.3	4.9
Mexico	17	1.2	1.1
Brazil	16	0.8	1.1
Western Europe	647	48.1	44.8
European Union (15)	605	42.9	41.9
Germany	133	9.7	9.2
United Kingdom	92	5.5	6.3
France	62	6.2	4.3
Italy	56	5.7	3.9
Central and Eastern Europe, the Baltic	59	3.0	4.1
Africa	37	3.3	2.6
Egypt	6	0.4	0.4
South Africa	5	0.4	0.4
Middle East	45	4.1	3.1
Israel	12	0.6	0.9
Asia	355	21.9	24.6
Japan	107	10.3	7.4
China	39	0.5	2.7
Korea, Rep. of	33	1.2	2.3
Hong Kong, China	25	1.4	1.7
Taipei, Chinese	24	1.7	1.6
India	203	0.7	1.6

Figure 14 Spatial distribution of world trade in merchandise and commercial services

The EU is the most important trade bloc with regard to exports and imports, accounting for well over a third of both movements. In rank order it is followed by NAFTA and ASEAN. However, according to both exports and imports, Japan is a larger trader than the ten ASEAN countries put together.

In recent decades trade in commercial services has increased considerably. However, in terms of total value it is still only about a quarter of that of merchandise trade. Again, the USA is the major importer and exporter. The UK ranks second in the export of services, ahead of Germany, France and Japan. Germany and Japan import a greater value of services than the UK.

2 The Role of the WTO

Trade is the most vital element in the growth of the global economy. In 1947 a group of 23 nations agreed to reduce tariffs on each other's exports under the General Agreement on Tariffs and Trade (GATT).

1947	Birth of the GATT, signed by 23 countries on 30 October at the Palais des Nations in Geneva.
1948	The GATT comes into force. First meeting of its members in Havana, Cuba.
1949	Second round of talks at Annecy, France. Some 5000 tariff cuts agreed to; ten new countries admitted.
1950–1	Third round at Torquay, England. Members exchange 8700 trade concessions and welcome four new countries.
1956	Fourth round at Geneva. Tariff cuts worth $1.2 trillion at today's prices.
1960–2	The Dillon round, named after US Under Secretary of State Douglas Dillon, who proposed the talks. A further 4400 tariff cuts.
1964–7	The Kennedy round. Many industrial tariffs halved. Signed by 50 countries. Code on dumping agreed to separately.
1973–9	The Tokyo round, involving 99 countries. First serious discussion of non-tariff trade barriers, such as subsidies and licensing requirements. Average tariff on manufactured goods in the nine biggest markets cut from 7% to 4.7%.
1986–93	The Uruguay round. Further cuts in industrial tariffs, export subsidies, licensing and customs valuation. First agreements on trade in services and intellectual property.
1995	Formation of World Trade Organisation with power to settle disputes between members.
1997	Agreements concluded on telecommunications services, information technology and financial services.
1999	Ministerial conference in Seattle in December 1999, which became a focus for opposition to 'globalisation'. Membership now 135 countries.
2001	In Nov. the Doha Summit launch a new round of free trade talks.
2002	1 January: China joins the WTO.

Figure 15 A GATT/WTO chronology

This was the first multilateral accord to lower trade barriers since Napoleonic times. Since the GATT was established there have been nine 'rounds' of global trade talks, of which the most recent, the Doha (Qatar) round, began in 2001 (Figure 15). A total of 142 member countries were represented at the WTO talks in Doha.

The most important recent development has been the creation of the World Trade Organisation (WTO) in 1995. Unlike its predecessor, the loosely organised GATT, the WTO was set up as a permanent organisation with far greater powers to arbitrate trade disputes. Figure 16 shows the benefits of the global trading system according to the WTO.

Although agreements have been difficult to broker at times, the overall success of GATT/WTO is undeniable: today average tariffs are only one-tenth of what they were when GATT came into force and world trade has been increasing at a much faster rate than GDP. However, in some areas protectionism is still alive and well, particularly in clothing, textiles and agriculture (Figure 17). In principle, every nation has an equal vote in the WTO. In practice, the rich world shuts the poor world out in key negotiations. In recent years agreements have become more and more difficult to reach, with some economists forecasting the stagnation or even the break-up of the WTO.

Relations between the USA and the EU have recently been soured by the so-called 'banana war', and by disagreements over hormone-treated beef, GM foods and steel (Figure 18). Leading agricultural exporters such as the USA, Australia and Argentina want a considerable reduction in barriers to trade for agricultural products. Although the EU is committed in principle to reducing agricultural support, it wants to move slowly arguing that farming merits special treatment because it is a 'multifunctional activity' that fulfils important social and environmental roles. Many developing countries have criticised the WTO for being too heavily influenced by the interests of the USA and the EU.

The WTO exists to promote free trade. Most countries in the world are members and most who are not want to join. The fundamental issue is; does free trade benefit all those concerned or is it a subtle way

1. The system helps promote peace
2. Disputes are handled constructively
3. Rules make life easier for all
4. Freer trade cuts the cost of living
5. It provides more choice of products and qualities
6. Trade raises incomes
7. Trade stimulates economic growth
8. The basic principles make life more efficient
9. Governments are shielded from lobbying
10. The system encourages good government

Figure 16 The benefits of the WTO trading system (according to the WTO)

The latest round of World Trade Organisation talks in Qatar proves that Western trade blocs need developing countries on-side to get a deal

Two years ago in Seattle, attempts to launch a new round collapsed, after developing countries walked out in protest at US threats to link trade with labour standards, a move many poor countries regard as a ruse to keep their goods out of western markets.

If Seattle was a watershed for the developing world, they came of age as a political force in Doha. Thanks to India's last-minute threat to walk out, they successfully resisted EU attempts to load the agenda with a range of complex new issues. They wrung some promises out of the west to open up vital markets, and won a major victory on the issue of drug patents and public health. But describing the results of Doha as a development agenda, as European ministers have done, is a little premature. Aside from the patent issue, the biggest victories for developing countries were defensive and the main concession was extra time to implement existing agreements.

Of the individual issues discussed at Doha, agriculture as always was the most contentious, pitting the EU against the rest of the world which wanted the EU to commit to eliminating its export subsidy system. After much last-minute wrangling, a compromise deal was reached. Negotiations will discuss phasing out export subsidies, but there was a figleaf for Europe: the words "without prejudging the outcome" were inserted.

In the argument over drug patents, India, Brazil and African countries were at odds with the EU and the US over whether existing WTO intellectual property rules allow enough flexibility for poor countries to buy cheaper generic drugs. The outcome was an outright victory for the developing countries which won a widespread exemption to patent rules in the interests of public health which should bring down the price of patented medicines.

After agriculture and the patent issues, the most pressing issue for developing countries were demands from the EU and the US that the WTO begin devising new global rules for competition, investment, government procurement and trade facilitation. When the Indian trade minister threatened to walk out, it was agreed no negotiations will begin on these issues until after the WTO's next ministerial meeting in two years' time and any country can veto the talks.

The environment was one of the few areas where the EU made significant gains, despite fierce opposition from developing countries which suspect the EU's real agenda is green protectionism. For the first time, the WTO will discuss the relationship between its rules and multilateral environmental agreements such as the Kyoto protocol on climate change.

One stark difference between Doha and previous WTO summits was that all the big trade blocs recognised that they needed to get poor countries on-side to get a deal, which could explain why the US was prepared to concede early on the patents issue.

The key to whether the Doha development agenda lives up to its billing is whether they carry on paying heed to developing countries' views once negotiations start in earnest in Geneva. A WTO round aimed at delivering for poor countries would focus on opening agriculture and textiles markets and on reducing tariff peaks on key developing country goods, which mean they pay four times as much to get their goods into western markets compared to developed economies.

As Tanzania's industry and trade minister, Iddi Mohamed Simba, put it, issues like agriculture and textiles may lose votes in Europe and the US, but "they are matters of life and death for us". The successful new trade round, he says, is far more important for poor countries than for rich ones. "We need the WTO far more than the US and the EU do."

Figure 17 Source: *The Guardian*, 23 November 2001

in which the rich nations exploit their poorer counterparts? Most critics of free trade accept that it does generate wealth but they deny that all countries benefit from it. For example, Barry Coates, Director of

Four months after the WTO launched a new round of global trade talks in Doha, the USA imposed tariffs of up to 30% on steel imports to protect its own fragile steel industry. More than 30 US steel producers went bankrupt between 1997 and 2002. Those that remained were considered to be inefficient and high cost compared to most of their foreign counterparts. Management consultants have largely put this down to the strength of the steel unions and their demands for high wages and health insurance. The crux of the problem is that world steel making capacity, estimated at between 900 million and 1000 million tonnes, is 20% higher than current demand. Although restructuring has already occurred, more is bound to happen both in the USA and in other parts of the world.

The reaction of America's trading partners was not difficult to predict. Trade unionists warned that the new trade barriers could result in 5000 job losses in Britain and 10,000 in the EU as a whole. The countries affected by the new tariffs argued that the USA was in breach of WTO rules. They also announced that they would demand compensation from the USA for the effect of the tariffs. However, as it could take up to two years for the WTO to reach a judgement, significant damage could be done in the intervening period to the steel industries of those nations affected. To its credit the EU stated that any retaliatory action would be within WTO rules. Overall this dispute was the last thing that the global steel industry, worth an estimated $500 billion, wanted.

Figure 18 Trade wars: steel

the World Development Movement wrote in the *Observer* (21 November 1999), 'Trickle down to the poor hasn't happened. In the past 20 years, the developing countries share of world trade has halved, income per person has fallen in 59 countries, and the number of people living on less than $1 a day has risen dramatically'. The non-governmental organisation Oxfam is a major critic of the way the present trading system operates. Figure 19 shows the main goals of its 'Make Trade Fair' campaign.

However, others would view the data available in a different way. Over the 1985–2000 period global inequality as measured by the Gini coefficient seems to have declined significantly (Figure 20). The main reason for this has been the rise in living standards in China and India. But what about the poor countries of Africa and elsewhere in the world? Supporters of the WTO say that it is scarcely credible to argue that the poverty of these countries is the result of globalisation since they are all outside the mainstream of free trade and economic globalisation. Critics of the WTO, on the other hand, say that the WTO and other international organisations should be paying more attention to the needs of these countries, making it easier for them to become more involved in, and gain tangible benefits from, the global economic system.

Critics of the WTO ask why it is that MEDCs have been given decades to adjust their economies to imports of textiles and agricultural products from LEDCs when the latter are pressurised to open their borders immediately to MEDCs banks, telecommunications

1. End the use of conditions attached to IMF–World Bank programmes that force poor countries to open their markets regardless of the impact.
2. Improve market access for poor countries and end the cycle of subsidised agricultural over-production and export dumping by rich countries.
3. Change WTO rules so that developing countries can protect domestic food production.
4. Create a new international commodities institution to promote diversification and end over-supply in order to raise prices for producers and give them a reasonable standard of living.
5. Change corporate practices so that companies pay fair prices.
6. Establish new intellectual property rules to ensure that poor countries are able to afford new technologies and basic medicines.
7. Prohibit rules that force governments to liberalise or privatise basic services that are vital for poverty reduction.
8. Democratise the WTO to give poor countries a stronger voice.

Figure 19 Oxfam's 'Make Trade Fair' campaign

Figure 20 Gini coefficients of worldwide income distribution

companies and other components of the service sector. The removal of tariffs can have a significant impact on a nation's domestic industries. For example, India has been very concerned about the impact of opening its markets to foreign imports (Figure 21).

Opposition to the WTO comes from a number of sources:

- Many LEDCs who feel that their concerns are largely ignored.
- Environmental groups concerned, for example, about a WTO ruling that failed to protect dolphins from tuna nets.
- Labour unions in some developed countries, notably the USA,

Since India was forced by a WTO ruling to accelerate the opening up of its markets, food imports have quadrupled. Large volumes of cheap, subsidised imports have flooded in from countries such as the USA, Malaysia and Thailand. The adverse impact has been considerable and includes the following:

- Prices and rural incomes have fallen sharply. The price paid for coconuts has dropped 80%, for coffee 60%, and pepper 45%.
- Foreign imports, mainly subsidised soya from the USA and palm oil from Malaysia have undercut local producers and have virtually wiped out the production of edible oil.

The new emphasis on exports, in order for India to compete in the world market, is also threatening rural livelihoods. For example, in Andhra Pradesh funding from the World Bank and the UK will encourage farm consolidation, mechanisation and modernisation. In this region it is expected that the proportion of people living on the land will fall from 70 to 40% by 2020.

Farmers, trade unionists and many others are against these trends, or at least the speed with which they are taking place. They are calling for the reintroduction of import controls, thus challenging the linchpin of the globalisation process – the lowering of trade barriers.

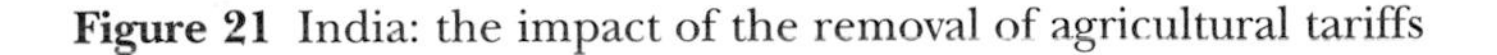

Figure 21 India: the impact of the removal of agricultural tariffs

concerned about: (a) the threat to their members jobs as traditional manufacturing filters down to LEDCs; and (b) violation of 'workers' rights' in developing countries.

3 The Terms of Trade

The most important element in the trade of any country is the terms on which it takes place. If countries rely on the export of commodities that are low in price and need to import items that are relatively high in price they need to export in large quantities to be able to afford a relatively low volume of imports. Many poor nations are *primary product dependent*, that is they rely on one or a small number of primary products to obtain foreign currency through export. The world market price of primary products is, in general, very low compared to manufactured goods and services. Also, the price of primary products is subject to considerable variation from year to year, making economic and social planning extremely difficult. In contrast, the manufacturing and service exports of the developed nations generally rise in price at a reasonably predictable rate resulting in a more regular income and less uncertainty for the rich countries of the world. The terms of trade for many developing countries are worse now than they were a decade ago. Thus, it is not surprising that so many nations are struggling to get out of poverty.

4 Africa: By-passed by the Benefits of Globalisation

An Oxfam report published in April 2002 stated that if Africa increased its share of world trade by just 1% it would earn an additional £49 billion a year – five times the amount it receives in aid. The World Bank has acknowledged that the benefits of globalisation are barely being passed on to sub-Saharan Africa and may actually have accentuated many of its problems.

The level of international aid given to Africa fell by 40% between 1990 and 1999. Although non-governmental organisations such as Christian Aid and Oxfam deplore this reduction they argue strongly that trade is the key to real development, being worth 20 times as much as aid. However, the trading situation of Africa will only improve if the trading relationship between MEDCs and LEDCs is made fairer to bring more benefits to the latter. In fact, Africa's share of world trade has fallen in recent decades. According to Oxfam, if sub-Saharan Africa had maintained its exports at the same level as 1980, its economy would be worth an extra $280 billion a year.

IMF–World Bank loans are usually conditional on African countries opening their markets. Historically, African trade barriers have been high but they have been reduced significantly in recent years. Although the situation varies across the continent, some countries such as Mali, Mozambique and Zambia are more open to trade than the EU and the USA. However, many countries complain that MEDCs, in particular the EU and the USA, are not implementing at home the free trade policies they expect African countries to follow. The high level of agricultural subsidies in the USA and the EU is a particular cause of concern, resulting in artificially cheap food flooding African markets. An example of the impact that such 'dumping' can have is the fate of the Ghanaian rice industry, which has collapsed in recent years as heavily subsidised US and Thai imports have undercut local producers. From being an exporter of rice, Ghana now imports £100 million of rice a year.

CASE STUDY: THE EMERGENCE OF CHINA AS A MAJOR TRADING NATION

Soon after the death of Mao in 1976, China's economic policy changed significantly. Mao's successor, Deng Xiaoping sought to end the relative isolation of China from the world economy and to imitate East Asia's export-led success. Economic growth increased by an average of over 10% a year and exports (by value) by 15% a year in the 1980s and 1990s. During this 20 year period the Chinese economy grew eight times bigger, and between 1990 and 1998 the number of Chinese living on less than a dollar a day fell by 150 million. Since the Chinese economy began to open up to the outside world in 1978 China's share of world trade has quadrupled. Under the traditional calculation of GDP, China ranks sixth in the world. But measured according to purchasing power parity (PPP), whereby the figures are adjusted to take account of price differences between countries, China ranks second in the world.

The most significant recent event in the history of the WTO has been Chinese entry into the organisation. On 15 November 1999 The USA and China agreed in principle to a deal that would allow China to join the WTO. Although there were other hurdles to get over, this was the most important obstacle to clear. The success of the USA–China negotiations was hailed as a triumph for the moderate, reformist policies of Chinese President Jiang Zemin. For the MEDCs in particular the main benefit of Chinese entry was:

- Its huge market potential. Car sales are taking off, exceeding 1 million in 2002 for the first time. China is now VW's biggest market outside Germany. China now imports more from the rest of Asia than does Japan, and consumes more copper and steel than the USA. As WTO membership opens China's markets to competition, its importance as a source of demand will grow.
- That China would be bound by WTO rules on a range of issues concerning production and trade.

The main concerns for China were:

- the problems caused by the new rules that China had signed up to as the country struggled to identify and specialise in fields of comparative advantage. In 1999 the World Bank estimated that up to a third of the 140 million workers employed in China's state-owned industries may be surplus to requirement.

China attracted a record $52.7 billion in *foreign direct investment (FDI)* in 2002, taking over from the USA as the world's biggest net recipient of FDI. The Chinese government expects to attract about $100 billion in FDI a year between 2006 and 2010. The Chinese government has made it easier for foreign companies to expand in China since the country joined the WTO. The major attraction to manufacturers is the cheap labour market where wages are less than 5% of those in the USA. The Chinese economy grew by 8% in 2002.

China now makes 60% of the world's bicycles and over half of the world's shoes. It accounts for 20% of the world's garment exports, with the prediction that this will rise to 50% in 2010 as quotas on imports are eliminated around the world. However, worries about Chinese goods swamping global markets seem to be exaggerated. China's share of world trade is still only 4%, with a trade surplus of about $30 billion (similar to Canada's). Although China is steadily producing more capital-intensive goods these are mainly destined for the domestic market where demand is rising rapidly.

China has considerably more control over its economy than most other countries. Its currency is not freely convertible. Thus, the country was not vulnerable to the speculation and resulting panic that affected so many of its neighbours in the late 1990s.

World trade: large container ships in Seattle harbour

5 Regional Trade Agreements

Regional trade agreements have proliferated in the last decade. In 1990 there were less than 25; by 1998 there were more than 90. The most notable of these are the European Union, NAFTA in North America, ASEAN in Asia and Mercosur in Latin America. The United Nations (1990) refer to such organisations as 'geographically discriminatory trading arrangements'. Nearly all of the WTO's members belong to at least one regional pact. All such arrangements have one unifying characteristic: the preferential terms that trade participants enjoy over non-participating countries. Although no regional group has as yet adopted rules contrary to those of the WTO, there are some concerns:

- Regional agreements can divert trade, inducing a country to import from a member of its trading bloc rather than from a cheaper supplier elsewhere.

- Regional groups might raise barriers against each other, creating protectionist blocks.
- Regional trade rules may complicate the establishment of new global regulations.

There is a growing consensus that international regionalism is on the ascendency. The EU, NAFTA and ASEAN+ (associated agreements with other countries) triad of regional trading arrangements dominate the world economy, accounting for 67% of all world trade. Whether the regional trade agreement trend causes the process of world trade liberalisation to falter in the future remains to be seen.

Apart from trade blocs there are a number of looser trade groupings aiming to foster the mutual interests of member countries. These include:

- The Asia-Pacific Economic Co-operation forum (APEC). Its 21 members border the Pacific Ocean and include Canada, the USA, Peru, Chile, Japan, China and Australia. The member countries have pledged to facilitate free trade.
- The Cairns Group of agricultural exporting nations was formed in 1986 to lobby for freer trade in agricultural products. Its members include Argentina, Brazil, Canada, New Zealand, Australia, the Philippines and South Africa.

6 Bilateral Free Trade Agreements

A recent headline in the *Financial Times* (19 November 2002) read 'As countries clamour for bilateral agreements (between two countries), the prospects of creating a truly open global economy recede'. The article argued that the willingness of governments to use trade pacts to cement diplomatic ties and forge alliances risks slowing down the momentum behind multilateral trade talks. Many economists argue that bilateral deals are a less effective route to trade liberalisation. Because trade within bilateral agreements is conducted on preferential terms, these pacts discriminate against third countries. The concern of many is that as more and more bilateral agreements are signed, the aim of a more seamless global economy will be hindered by a patchwork quilt of agreements. Because of this the EU, which is party to a number of bilateral agreements, plans to launch no new negotiations until after the Doha round of WTO talks ends. The USA aims to ensure the provisions of its future FTAs are uniform by adopting a standard template. With the Doha round facing tough challenges, and its end-2004 deadline looking increasingly hard to attain, some countries may decide to concentrate their efforts on smaller deals that offer the prospect of more immediate results.

7 Trade Blocs

A trade bloc is a group of countries that share trade agreements between each other. Since the Second World War there are many examples of groups of countries joining together to stimulate trade between themselves and to obtain other benefits from economic co-operation. The following forms of increasing economic integration between countries can be recognised:

- *Free trade areas* – members abolish tariffs and quotas on trade between themselves but maintain independent restrictions on imports from non-member countries. NAFTA is an example of a free trade area.
- *Customs unions* – this is a closer form of economic integration. Besides free trade between member nations, all members are obliged to operate a common external tariff on imports from non-member countries. Mercosur, established on 1 January 1995, is a customs union joining Brazil, Paraguay, Uruguay and Argentina in a single market of over 200 million people.
- *Common markets* – these are customs unions, which, in addition to free trade in goods and services, also allow the free movement of labour and capital.
- *Economic unions* – these organisations have all the characteristics of a common market but also require members to adopt common economic policies on such matters as agriculture, transport, industry and regional policy. The EU is an example of an economic

El Camino de Santiago; an example of European Union funding in Spain union, although it must be remembered that its present high level of economic integration was achieved in a number of stages. When Britain joined in 1973 the organisation could best be described as a common market. The increasing level of integration has been marked by changes in the name of the organisation. Initially known as the European Economic Community, it later became the European Community and finally, from November 1993, the European Union.

8 The Association of South-east Asian Nations (ASEAN)

ASEAN (developed out of the old SEATO anti-communist grouping) was established in 1967 by the five original member countries: Indonesia, Malaysia, the Philippines, Singapore and Thailand. Brunei joined in 1984, Vietnam in 1995, Laos and Myanmar, in 1997, and Cambodia in 1999. The ASEAN area, one of the largest regional markets in the world, has a population of around 500 million with a combined GDP of $737 billion. The stated objectives of ASEAN are:

(a) To accelerate the economic growth, social progress and cultural development of the region.
(b) To promote regional peace and stability.

When ASEAN was established trade among its members was very limited. The main stages in economic co-operation were:

- The Preferential Trade Arrangement of 1977, which set tariff preferences for trade among ASEAN economies.
- In 1987 an Enhanced PTA Programme was adopted at the third ASEAN summit in Manila further increasing intra-ASEAN trade.
- In 1992 the Framework Agreement on Enhancing Economic Co-operation was adopted at the Fourth ASEAN Summit in Singapore. This included the objective of establishing an ASEAN Free Trade Area or AFTA.
- The Fifth ASEAN Summit in 1995 adopted the Agenda for Greater Economic Integration, which included the acceleration of the timetable for the realisation of AFTA from the original 15-year timetable to ten years.
- In 1997 the member countries adopted the ASEAN Vision 2020 aimed at forging closer economic integration within the region.
- The 1998 Hanoi Plan of Action became the first of a series of plans of action leading up to the realisation of the ASEAN vision.

Within three years of the launching of AFTA, exports among ASEAN countries grew from $43 billion in 1993 to $80 billion in 1996. Trade between the member countries as a percentage of total exports increased from 17.35% in 1980 to 22.69% in 2000. Today, ASEAN economic co-operation covers trade, investment, industry, services, finance, agriculture, forestry, energy, transportation and communication, intellectual property, and tourism.

ASEAN has held regular talks with China, Japan and South Korea with a view to creating the world's largest free trade zone, which could be in operation by the end of the decade. A full East Asian trading bloc covering more than two billion people would be a formidable player on the world stage. At a meeting with the ASEAN nations in November 2002, the Japanese prime minister Junichiro Koizumi stated 'This region can be the economic engine of the world'. For China, membership of ASEAN will strengthen its regional influence. For the existing members it will offer a new channel for diplomatic dialogue and preferential access to a vast market.

CASE STUDY: THE NORTH AMERICAN FREE TRADE AGREEMENT (NAFTA)

NAFTA came into effect on 1 January 1994 with the objective of eliminating most tariffs and other restrictions on free trade and investment between the United States, Canada and Mexico by the year 2003; remaining tariffs will be removed by 2008.

Although the idea of a North American trade bloc had been around for some time the formation of NAFTA in the 1990s was hastened by three factors: the ever-increasing economic challenge from western Europe and Asia; the completion of the internal market of the European Union (EU) and the establishment of the European Economic Area (EEA) in 1993; and growing concern that nations left outside trade blocs would be commercially disadvantaged.

(a) Stages of Development

The first significant move towards a North American trade bloc was the signing of the Canada–United States Automotive Products Trade Agreement (Auto Pact) in 1965. In 1988 the two countries extended their relationship with the establishment of the comprehensive Canada–United States Free Trade Agreement (FTA or CUSTA). In 1990 Mexico formally requested a free trade relationship with its northern neighbours and after four years of intense negotiation NAFTA was approved by the governments of all three countries. In effect, the terms of the 1988 FTA were extended to include Mexico, whose economy was then less than 5% the size of those of the United States and Canada combined. This established a unique relationship between a relatively poor Third World nation and two of the world's richest countries. Never before had a trade bloc included both members of the developed and developing worlds. When NAFTA was established in 1994 its 390 million consumers with a combined GDP of over $7.6 trillion vied with the EEA (the EU and Iceland, Norway and Liechtenstein) to become the world's largest trade bloc.

(b) The NAFTA Agreement

The 1994 agreement planned for all tariffs on goods qualifying as North American to be phased out within ten years, although special rules applied to key sectors such as energy, agriculture, textiles and clothing. Trade in services would also be facilitated while other provisions would give relief or protection to 'sensitive industries' (e.g. some agricultural products like US sugar were given protection for 15 years) and technical and environmental standards. To lessen American concerns about Mexico, special 'supplemental agreements' or 'side deals' were added to NAFTA. These provide for annual reviews of 'import surges' or dumping of products and, if necessary, a penalty: a resumption of pre-NAFTA tariffs, usually for three years. In addition, three-country commissions monitor the enforcement of NAFTA-related national laws. Here any two nations can instigate an investigation of suspected breaches of regulations concerning environmental standards, health and safety in the workplace, minimum wages and child labour. Fines of up to $20 million can be imposed.

(c) Differences between NAFTA and the EU

The objectives of NAFTA differ from the EU's Maastricht Treaty (1993) in a number of significant ways. The NAFTA agreement is limited to trade only and thus does not:

- permit free movement of labour
- attempt to redistribute wealth to poorer regions within its boundaries
- seek to establish a common currency
- seek political union
- aim to establish a customs union with common external tariffs
- affect existing border controls.

However, like the EU, NAFTA may allow other countries to join providing all current members agree. The objective of NAFTA's supporters is that the whole of Latin America will follow Mexico's example and eventually join. The establishment of Mercosur on 1 January 1995 was a major stage in this process. The members of this customs union are Brazil, Argentina, Paraguay and Uruguay. Mercosur is negotiating with the Andean Pact countries (Bolivia, Peru, Ecuador, Colombia and Venezuela) and Chile for the creation of a South American Free Trade Area as an important building block towards the establishment of the American Free

Trade Area (AFTA) incorporating all the countries of North, Central and South America by the year 2005, as agreed during the 'Americas Summit' in Miami in December 1994. This idea of a trade bloc covering all of the Americas was reaffirmed by the Quebec Declaration in April 2001.

Not everyone is positive about the proposed FTAA. President da Silva of Brazil, when still in opposition, called the FTAA an 'attempt at annexing Latin America' by the United States. In September 2002 Brazil's catholic bishops organised a 'plebiscite', inviting the public to express opposition to the proposed accord. Some Brazilians argue that the WTO is a better forum than the FTAA in which to achieve free trade.

(d) The Impact on the USA

The arguments for and against NAFTA have been debated most fiercely in the United States. Within the USA the strongest proponents of NAFTA have been multinational corporations, economists and key political figures such as ex-President Clinton. The classical economic argument, following Ricardo's *Theory of Comparative Advantage*, is that all three countries would be better off with free trade as they would specialise increasingly in what they are best at.

Trade unions and environmental groups have led the argument against NAFTA. The major debate in the USA focuses on the issue of trade with Mexico. Trade unions have long feared that free trade with Mexico would result in wage and benefit reductions if US firms were to remain competitive against cheap Mexican labour. They also foresaw US companies moving to Mexico to take direct advantage of lower wage rates as well as being attracted south of the border by less demanding environmental legislation. In 1999 the average factory worker in the USA earned more than eight times as much as his Mexican counterpart.

Many environmental groups such as the powerful Sierra Club have been strong critics of NAFTA. They envisaged more severe environmental degradation in Mexico where environmental laws are lax and often unenforced. The Sierra Club also predicted that US environmental legislation would be watered down in the name of staying competitive with Mexico.

Transnational corporations that have moved operations to Mexico have, as expected, reaped higher profits. However, the trade deficit with Mexico has increased with no signs of the situation being reversed. But the predictions of an almost immediate adverse impact on the US economy as a whole did not materialise as growth remained strong in the USA throughout the 1990s. Nevertheless, a significant number of American workers are less well off now than before the advent of NAFTA. Although the overall unemployment rate remained low into the new century, there is clear evidence that many higher paid workers have been forced into lower paid jobs as US companies have transferred manufacturing operations south of the border. Some companies have also used threats to move to Mexico when negotiating with workers.

Critics of NAFTA within the USA frequently cite the growth of the merchandise trade deficit as evidence for their point of view. However, NAFTA trade only accounts for 16% of the overall merchandise trade deficit.

Public Citizen, a pressure group opposed to NAFTA, has attacked the organisation over a wide range of issues, calling it a 'trade agreement from hell'. A recent publication (6 February 2001) entitled *Grave Danger Posed by Unsafe Mexican Trucks* argued that the USA should maintain limited access to its highways. A NAFTA tribunal is expected to reject US concerns about truck safety and order the USA to permit Mexican trucks to have full access to American highways. Currently, Mexican trucks are confined to a narrow, 20-mile commercial zone near the border. Among the concerns voiced by the USA were:

- Mexico does not limit the time drivers spend behind the wheel.
- Mexico's hazardous materials control system is much more lax than the US system.
- In 2000 Mexican truck carriers were found to be three times more likely to have safety deficiencies than US carriers.

(e) The Impact on Canada

Although there are significant exceptions, most organisations and individuals in Canada hold favourable views about NAFTA because the benefits of the agreement seem very clear cut. Merchandise trade with the USA increased 80% in the first five years of NAFTA, reaching $475 billion in 1998. During the same period merchandise trade with Mexico doubled to reach $9 billion.

- Canadian exports to the USA and Mexico increased 80% and 65%, respectively, in the first five years of NAFTA, reaching $271.5 billion and $1.4 billion.
- US investment in Canada reached $147.3 billion in 1998, up 63% from 1993. Investment from Mexico reached $464 million in 1998, tripling from 1993.
- More than one million new jobs have been created in Canada since the start of 1994.
- In 1998, 68% of foreign direct investment into Canada came from the USA and Mexico.

Canada's trade with its NAFTA partners has been growing much faster than its trade with other countries. Although Mexico's

The Waterfront, Vancouver

trade surplus with Canada rose considerably through the 1990s, Canada's surplus with the USA is on a much larger scale. Canada has always been conscious of the limited size of its domestic market and saw huge benefits of having open access to the USA and Mexico. As Canada is distant from Mexico it has not experienced some of the difficulties that have arisen between the USA and Mexico because of their common border. However, environmental groups in Canada have voiced similar concerns to those raised in the USA.

(f) The Impact on Mexico

Supporters of NAFTA in Mexico say the new market is forcing Mexican companies to adopt higher foreign standards and business practices. Such a process will gradually improve the competitiveness of Mexican business. As the Mexican economy is locked into the economies of the United States and Canada it makes it impossible for the country to revert to the disastrous protectionist policies of the past. In fact, Mexico has numerous trade agreements with countries other than its northern neighbours. Mexico trades at reduced or zero tariff with over 60% of the world, measured by GDP. It is the only country other than Israel that has free-trade agreements with both the United States and the European Union. Of all the Latin American nations, Mexico holds the greatest promise of development as a consumer market. Sound macroeconomic policies implemented by the

Zedillo government over the past five years have brought the country from profound economic crisis to a relatively solid economic footing.

However, not all Mexicans are so convinced about the merits of NAFTA. They argue that Mexico has swapped one kind of trade dependence for another. In the early 1980s oil dominated the country's economy, accounting for two-thirds of exports. Although the economy is more diversified now, over 88% of exports go to the United States, up from 78% at the beginning of the 1990s. However, while total US imports grew by 110.3% during the 1990s, Mexican exports to the USA grew by 252.5%, implying an increase in Mexico's share of US imports from 6.45% to 10.75%.The gain in market share during this period is equivalent to $44 billion in exports. Mexican exports have increased not only because of US demand but also because of Mexican penetration of the US market. This reflects: (a) the decline in barriers to trade; and (b) the improved quality of Mexican goods. Mexico has achieved significant market penetration in food and live animals, beverages and tobacco, machinery and transport equipment, and miscellaneous manufactured articles.

Bilateral trade between the USA and Mexico, which stood at $82 billion in 1992, rose to about $200 billion in 1999. However, while in many ways Mexico's success in the US market is good news, any downturn in the US economy is bound to have an adverse knock-on effect south of the border. A common expression in Mexican business circles is 'When the United States catches a cold, Mexico gets pneumonia'.

Another criticism is that the Mexican government did too little to prepare the country for such a significant change. Mexican farmers have been particularly hard hit. The vast majority of farm plots are less than 10 hectares and operate with very modest equipment. Before NAFTA they were protected by import tariffs and government-guaranteed prices. Now they have to compete with large-scale high-technology American and Canadian agribusiness. The impact on the landscape of the north-western state of Sonora has been startling. The state used to be known as Mexico's breadbasket because wheat dominated agricultural production here. Now, due to the impact of cheaper wheat from the United States and Canada, farmers have turned to nuts, peaches, asparagus, chickpeas, olives, cucumber, watermelon and jalapeno chilli. Most of production of the new crops is for export. Although change was already underway before 1994, the advent of NAFTA speeded up the process considerably. Corn growers have also been hit hard because the government has allowed more corn to enter the country duty-free than NAFTA specifies. Corn is Mexico's staple crop, covering about half the cultivated area.

At present Mexico must import more than 50% of its milk supply. Most comes from the USA, along with many other agricultural products. However, other Latin American countries are also gaining a foothold in Mexico's food market by means of free trade agreements.

Industry too has had its problems. Critics of the government claim that Mexico does not have an industrial policy apart from promoting *maquiladoras*. These are factories that import materials or parts to make goods for re-export. Maquiladoras existed long before 1994 but they have increased greatly in number since NAFTA came into effect. Although the jobs they provide are important to the economy, less than 3% of the maqiuladoras' input is produced locally. Thus, while rising exports have been helpful in reducing the trade deficit, they have not done much for the rest of the economy. While exports have been growing by an average of 10.9% a year in the 1980s and 1990s, output for the domestic market has risen by an average of only 3.5%.

The export boom has had a very uneven impact in Mexico. Employment in the maquiladoras is now 1.3 million compared with 546,000 when NAFTA began. However, the vast majority of these factories are close to the border with the USA, the main destination of the finished products.

An ever-increasing number of Asian and European companies have established plants in Mexico in order to gain access to Mexico's NAFTA trade partners. Mexico is being used as a springboard to the USA and Canada in the same way that many foreign companies base operations in the UK to gain open access to the whole of the EU. At the same time, many US companies manufacture products in Mexico that are destined for the latter's trade partners throughout Latin America.

A clear-cut assessment of NAFTA membership on Mexico is difficult because many of Mexico's trade liberalisation policies were in effect before NAFTA began, prompted by Mexico's membership of GATT and its ongoing domestic reforms.

Conclusion

Opinion about NAFTA remains divided, particularly in the USA. Some analysts say its too early to judge NAFTA's impact, partly because many of its provisions have yet to take effect. Also, trade agreements are directly influenced by macroeconomic changes in individual countries and globally, such as changes in income and exchange rates. A reasonable statement about NAFTA would be that neither the critics' worst fears nor the supporters' rosiest forecasts has materialised.

Summary

- Barriers to trade have fallen significantly under the auspices of first the GATT and then its successor the WTO.
- A significant number of countries feel that the WTO trading system operates in a manner which is unfair to LEDCs.
- The most important recent WTO event has been the accession of China in 2002.
- From time to time 'trade wars' erupt between countries or trade blocs when one side or the other, or both, feel that the trading system is operating unfairly against them.
- The rapid removal of tariff barriers can have a major impact on certain sectors of a country's economy.
- Regional trade agreements have proliferated in the last decade or so.
- The EU, NAFTA and ASEAN+ triad dominate the world economy.
- There are some concerns that the growth of regional trade agreements and bilateral agreements will slow down the WTO's attempts to reduce further global trade barriers.

Questions

1. Study Figure 14.
   - **(a)** Describe the pattern of merchandise exports and imports for 2001.
   - **(b)** To what extent has the pattern changed since 1990?
   - **(c)** Summarise the pattern of exports and imports for commercial services in 2001.
   - **(d)** Suggest why the UK is ranked second for the export of commercial services.
2. **(a)** What is the function of the WTO?
   - **(b)** Why are many LEDCs critical of the way the WTO operates?
   - **(c)** Discuss some of the issues that have caused disagreements among WTO members.
3. **(a)** What is: (1) a regional trade agreement; and (2) a bilateral trade agreement?
   - **(b)** Why might such agreements make further progress at the WTO difficult?
   - **(c)** Explain the differences between the various degrees of integration between countries.
4. **(a)** Describe the development of the North American Free Trade Agreement.
   - **(b)** Why did the USA, Canada and Mexico form a trade bloc?
   - **(c)** Discuss the main differences between NAFTA and the EU.
   - **(d)** Examine the impact of NAFTA on the USA, Canada and Mexico.

3 Globalisation and Changing Economic Activity

KEY WORDS

Clark-Fisher model: a model showing the changing relative importance of the sectors of employment over time.
Business cycle: the regular pattern of upturns and downturns in demand and output within an economy that tend to repeat themselves every five years or so.
Kondratiev long waves: 50-year cycles of economic upturn and downturn with each wave being associated with significant technological change.

1 The Information and Communications Revolution

The transformation towards a knowledge-based global economy has been driven by the growing role of *information and communication technologies* (ICT). Plunging communication costs have created new ways of organising firms at a global level.

The global telephone network has grown nearly tenfold in the past 40 years, reaching almost a billion lines by 2000. Teledensity, the number of lines per 100 people, has quadrupled since 1960, although a quarter of the world's nations still have a teledensity below one. However, the speed of network growth continues to increase rapidly.

Of enormous significance are the new information pipelines linking the players of greatest importance in the global economy. The transoceanic copper wires that made communications possible in the pre-satellite era are being replaced by arrays of sophisticated fibre optics capable of carrying huge amounts of data. Where these prime conduits for the technological revolution are routed is of great significance for change in the future. The most important of the recently constructed links are:

- The FLAG (Fiberoptic Link Around the Globe) project, a $1.5 billion, 28,000 km underwater cable snaking its way across the ocean floor from Britain to Japan. It offers uninterrupted data traffic between Europe and Asia, traffic that was previously routed through the USA. The magazine *Time* stated in 1997 'It is akin to opening a new navigation route that will link 75% of the world's population. Its five gigabits of information a second will allow for a huge increase in electronic traffic'.
- The SEA-ME-WE 3 cable, approved in January 1997 by the 70 countries involved in its construction. The $1.73 billion project

stretches 38,000 km and connects western Europe, the Middle East and South-east Asia with fibre-optic technology capable of carrying 120,000 simultaneous phone conversations.

Fibre-optic cables have a number of advantages over satellite transmission. Voice and data traffic is faster, cheaper, more reliable and subject to less interference. However, the submarine pipelines are vulnerable to both physical and human attack.

The cost of computer processing power has been falling by an average of 30% a year in real terms over the past couple of decades. With the costs of communication and computing falling rapidly, the natural barriers of time and space that separate national markets have been tumbling too.

2 Models of Economic Change

Global economic activity clearly changes over time. A number of models have been produced to explain the changes that occur.

(a) The Clark-Fisher Model

This model (Figure 22) shows how, over time, the relative importance of the sectors of employment change. In the pre-industrial phase the primary sector dominates as it did in Britain prior to the Industrial Revolution. The proportion of the workforce employed in manufacturing increases as a country's economy diversifies away from heavy reliance on the primary sector. This usually begins with low-technology industries based on locally available raw materials. Over time the

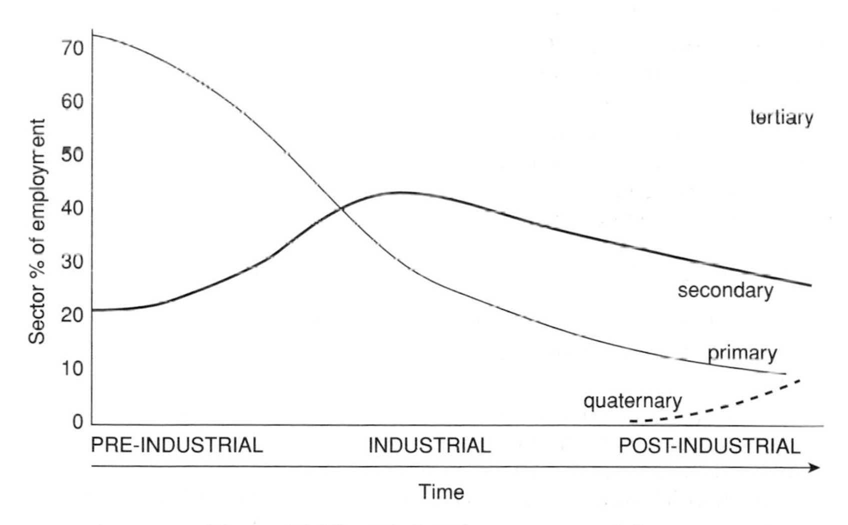

Figure 22 The Clark-Fisher sector model

manufacturing base widens. To support the increasing number of industries and the demands of a more affluent population, a growing range of services is required, leading to the expansion of tertiary employment.

Changes in the sectors of employment can change quite quickly. In 1900, 40% of employment in the USA was in the primary sector. However, the mechanisation of farming, mining, forestry and fishing drastically reduced the demand for labour in these industries. As these jobs disappeared people moved to urban areas where most secondary, tertiary and quaternary employment is located. Less than 4% of employment in the USA is now in the primary sector.

Human labour is steadily being replaced in manufacturing too. In more and more factories, robots and other advanced machinery handle assembly line jobs that once employed large numbers of people. In 1950, the same number of Americans were employed in manufacturing as in services. By 1980, two-thirds were working in services.

The tertiary and quaternary sectors are also changing. In banking, insurance and many other types of business, computer networks have reduced the number of people required. But elsewhere service employment is rising such as in health, education and tourism.

Globalisation has been a significant factor in the changing importance of the sectors of employment in many countries. Such change has occurred most rapidly in the Newly Industrialised Countries, the success story of globalisation. In the least developed countries, those nations that have been largely by-passed by the rapidly evolving global economy, there is still heavy reliance on the primary sector.

(b) The Business Cycle

Economic growth and the integration of the global economy is not a continuously smooth process. It tends to operate in fits and starts. Two timescales have been identified in this irregular process – the business cycle, which is relatively short term, and Kondratiev long waves, which operate over a considerably greater time frame.

The business cycle is the regular pattern of upturns and downturns in demand and output within the economy that tend to repeat themselves every five years or so. Over the years most governments have pursued policies to try to smooth out these 'booms' and 'busts' with limited effect. Although the causes of these undulations in the economy are not fully understood, significant contributory factors to the business cycle are:

- The bunching of investment spending, which leaves an 'investment shadow' period after it.
- Business confidence or lack of it. If the outlook is promising many firms expand investment, at the same time contributing to an

economic upturn. Conversely, when the outlook for the economy is poor a large number of companies postpone investment plans.
- Government policies that attempt to increase economic growth in the period prior to elections. This may lead to inflation and thus the need to constrain the economy after an election.

If governments do, in fact, succeed in smoothing out the familiar upturns and downturns in national economies the process of globalisation should increase in pace.

(c) Kondratiev Long Waves

This is the theory, generally associated with the work of the Russian economist ND Kondratiev in the 1920s, that in addition to the five year business or trade cycle, there exists a 50-year cycle of economic upturn and downturn (Figure 23). Each wave, which is associated with significant technological change, has four phases: prosperity, recession, depression and recovery. As Figure 23 shows, each successive Kondratiev wave has a specific geography. The first wave was based on the early mechanisation of the industrial revolution, being largely confined to three countries. The fifth wave, which seems to have begun in the 1980s and 1990s and which is strongly based on information technology, involves a much wider range of nations. The latter point is not surprising due to the growing complexity of the linkages in the global economy.

3 The Globalisation of Primary Activities

(a) Agriculture

Farming and food production around the world is becoming increasingly dominated by large biotechnology companies, food brokers and huge industrial farms. The term agri-business is often used when describing the scale of the global food industry. The result is a complex movement of food products around the world. The food products, both fresh and processed, available in a typical UK supermarket have a much wider global reach than they did 20 years ago. Peoples' expectations in terms of variety and quality of food have never been higher, which has increased the pressure that the large food retailers have placed on suppliers. Large supermarkets and other bulk purchasers are generally able to switch suppliers faster than ever before, in some cases buying from a different continent. The falling real cost of transport has been an important factor in this trend.

However, there has been an increasing reaction to high input farming as more and more people have become concerned about the use of fertilisers, pesticides, herbicides and other high investment farming practices. The main evidence of this concern is the growth of

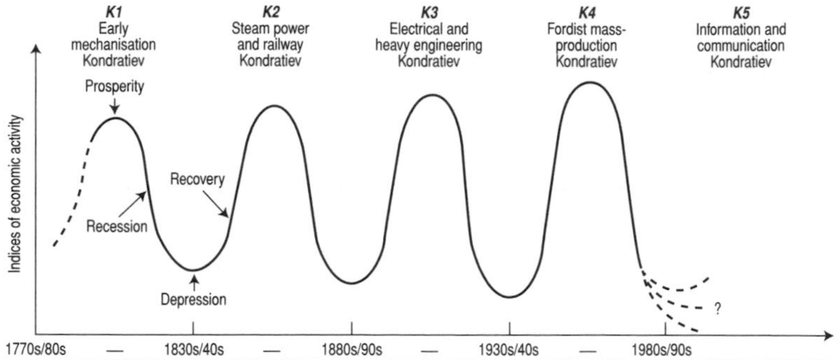

K1 Early mechanisation Kondratiev

K2 Steam power and railway Kondratiev

K3 Electrical and heavy engineering Kondratiev

K4 Fordist mass-production Kondratiev

K5 Information and communication Kondratiev

Prosperity

Recovery

Recession

Depression

Indices of economic activity

1770s/80s — 1830s/40s — 1880s/90s — 1930s/40s — 1980s/90s

	1770s/80s — 1830s/40s	1830s/40s — 1880s/90s	1880s/90s — 1930s/40s	1930s/40s — 1980s/90s	1980s/90s —
Main 'carrier' branches	Textiles Textile chemicals Textile machinery Iron working/castings Water power Potteries	Steam engines Steamships Machine tools Iron and steel Railway equipment	Electrical engineering Electrical machinery Cable and wire Heavy engineering/ armaments Steel ships Heavy chemicals Synthetic dyestuffs	Automobiles Trucks Tractors Tanks Aircraft Consumer durables Process plant Synthetic materials Petrochemicals	Computers Electronic capital goods Software Telecommunications Optical fibres Robotics Ceramics Data banks Information services
Infrastructure	Trunk canals Turnpike roads	Railways Shipping	Electricity supply and distribution	Highways Airports/airlines	Digital networks Satelites
Limitations of previous techno-economic paradigms. Solutions offered by new paradigms	Limitations of scale, process control and mechanisation in domestic 'putting out' system; of hand-operated tools and processes. Solutions offering prospects of greater producitivity and profitability through mechanisation and factory organisation in leading industries	Limitations of water power *re*: inflexibility of location, scale of production, reliability and range of applications restricting development of mechanisation and factory production to the economy as a whole. Largely overcome by steam engine and new transport system	Limitations of iron as an engineering material (strength, durability, precision, etc.) partly overcome by universal availability of cheap steel and alloys. Limitations of inflexible belts, pulleys, etc., driven by one large steam engine overcome by unit and group drive for electrical machinery, overhead cranes, power tools permitting vastly improved layout and capital saving. Standardization facilitating worldwide operations	Limitations of scale of batch production overcome by flow processes and assembly-line production techniques, full standardization of components and materials and abundant cheap energy. New patterns of industrial location and urban development through speed and flexibility of automobile and air transport. Further cheapening of mass-consumption products	Diseconomies of scale and inflexibility of dedicated assembly-line and process plant partly overcome by flexible manufacturing systems,. 'networking' and 'economies of scope'. Limitations of energy and materials intensity partly overcome by electronic control systems and components. Limitations of hierarchical departmentalisation overcome by 'systemation', networking' and integration of design, production and marketing
Organisation of firms and forms of co-operation and competition	Individual entrepreneurs and small firms (<100 employees) competition. Partnership structure facilitates co-operation of technical innovators and financial managers. Local capital and individual wealth	High-noon of small-firm competition, but larger firms now employing thousands rather than hundreds. As firms and markets grow, limited liability and joint-stock company permit new pattern of investment, risk-taking annd ownership	Emergence of giant firms, cartels, trusts, mergers. Monopoly and oligopoly become typical. Regulation or state ownership of 'natural' monopolies and public utilities. Concentration of banking and 'finance-capital'. Emergence of specialised 'middle management' in large firms.	Oligopolistic competition. Transnational corporations based on direct foreign investment and multiplant locations. Competitive subcontracting on 'arm's length' basis or vertical integration. Increasing concentration, divisionalisation and hierarchical control. 'Techno-structure' in large corporations	'Networks' of large and small firms based increasingly on computer networks and close co-operation in technology, quality control, training, investment planning and production planning ('just-in-time'), etc. 'Keiretsu' and similar structures offering internal capital markets
Geographical focus	Britain, France, Belgium	Britain, France, Belgium, Germany, USA	Germany, USA, Britain, France, Belgium, The Netherlands, Switzerland	USA, Germany, other EEC, Japan, Switzerland, Sweden, other EFTA, Canada, Australia	Japan, USA, Germany, other EEC and EFTA, Sweden, Taiwan, Korea, Caanda, Australia

Figure 23 Kondratiev long waves (P Dicken)

the organic food market and the increasing sale of food resulting from more 'gentle' farming practices. The Fair Trade movement is at the forefront of attempts to safeguard the position of small producers.

Large agricultural companies are continually seeking to produce new products, increase market share and lower costs in order to increase their competitiveness. One of the most worrying developments for many people is the spread of genetically modified farming. In 2000 the UK began a three-year trial of GM oilseed rape in fields

Pros	Cons
• Higher yields for farmers • Better resistance to virus and fungus attack so reducing the need to use chemicals • Reduce the need for spraying herbicides. Because the crop is weedkiller resistant the weeds need only be sprayed when they become a nuisance. • Genes could add beneficial food additives to plants to reduce heart disease or vitamin deficiency. • Production of edible vaccines in tomatoes or bananas, such as hepatitis B in developing countries. • Drought and salt resistant varieties for countries with poor land and/or low rainfall, particularly where bad irrigation has damaged the soil. • GM trees would contain fewer contaminates and grow faster so paper making would cause less pollution.	• Food safety is not guaranteed and not tested. Biotechnology could alter characteristics of staple crops such as potatoes and wheat. • Antibiotic marker genes could add to problems of resistance of bacteria to antibiotics in human and veterinary medicines. • No benefit to the consumer in reduced prices. • GM contamination of organic and conventional crops seen as inevitable in Britain. • Farmers who grow and sell GM-free seeds or save their own seeds face them being contaminated by cross pollination from GM crops. • Honey will be contaminated with GM pollen. • Superweeds with herbicide resistant genes – already a problem in Canada. • Weedkillers used on GM crops could kill all plants except the crop leaving nothing for insects and birds to feed on, damaging biodiversity.

Figure 24 Advantages and disadvantages of GM crops

across the country, with significant opposition from environmental groups. Figure 24 summarises the advantages and disadvantages of GM crops.

An increasing percentage of agricultural land in the developing world has come under the control of outside influences either directly or indirectly. Direct ownership by TNCs invariably means the cultivation of crops for export at the expense of food production for the domestic population. But even when farms are not foreign owned, IMF policies may dictate that land is used to produce for the export market. This may not always be a bad thing but if a country is undergoing food shortages at the same time it is an ironic situation. There is a growing realisation that the modes of production, processing, distribution and consumption that prevail, because in the short-to-medium term they are the most profitable, are not necessarily the most healthy or the most environmentally sustainable.

Between 1961 and 1999 there was a fourfold increase in the amount of food exported. This considerable increase in 'food miles' has a significant environmental impact. Trade-related transportation is one of the fastest growing sources of greenhouse gas emissions. Also, the long-distance movement of food and feed increases the risk of the spread of diseases such as foot and mouth and mad cow disease.

(b) Fishing

In recent decades as fish stocks have dwindled in more and more fishing areas the nations with the largest fishing fleets have looked further and further afield to maintain their catches. Such action has often had a hugely detrimental effect on small-scale local fishermen. A good example of this conflict of interests is along the coast of western Africa where factory ships from Europe and elsewhere are taking so many of the available fish that very little is left for the local industry. The largest of the foreign invaders is probably the Irish-owned Atlantic Dawn, which weighs 14,000 tons, is 145 metres long and carries a crew of 100. Taking in up to 400 tons of fish a day, the Atlantic Dawn can bring in a haul worth more than $2 million on the wholesale market.

In spite of the devastation this has caused to local fishing industries it is all perfectly legal. For decades Europe has leased fishing rights from Senegal, Mauritania and other West African nations. This is done because nations have legal entitlement under international law to all fish within 200 miles of their coastlines. However, critics of these leasing agreements argue that they are manifestly unfair because:

Small, local fishermen have suffered from over-fishing by MEDCs

- The contracts set no upper limit on the amount of fish that EU boats can catch.
- The countries of West Africa are so poor that they feel they have little choice but to accept the contract payments from the EU. For example, the EU will pay Senegal $500 million for the rights to fish its waters from 2001 to 2006.

In 1997 Senegal produced 453,000 tons of fish, but by 2000 production had declined to 330,000 tons. This has had a considerable effect on employment and on diet. In West Africa fish is an important source of dietary protein. Critics of the EU's exploitation of these waters see it as a form of modern colonialism.

(c) Energy and Minerals

International trade in minerals has been occurring for centuries but the level of trade today and the complexity of the trading networks is of a new order compared to anything that has gone before. For example, huge new gas pipeline networks have linked Russian gas fields in Siberia with consumers in western Europe and similar projects have linked together countries in South America. The electricity grids in some countries are now linked, as for example between Britain and France. As supplies of fuels and minerals have declined in traditional mining areas, the large TNCs controlling the trade in these commodities have looked increasingly to frontier and offshore areas.

4 The Globalisation of Manufacturing

North America, western Europe and Japan account for almost four-fifths of world manufacturing production. Although this is a highly clustered situation, it was much more so in the past. In 1953 95% of manufacturing was concentrated in the industrialised economies. The decentralisation that has occurred over the last four decades or so has largely been the result of investment by TNCs in those developing countries able to take on manufacturing tasks at a competitive price. Over time the scale of TNC investment in the developing world spawned three generations of newly industrialised country. The most dynamic NICs have created large companies of their own, which are locating factories in developed nations such as Britain. The globalisation of manufacturing will be examined in more detail in Chapters 5 and 6.

5 The Globalisation of Services

As a result of a combination of globalisation and informationalisation, the production of services has become increasingly detached from that of the production of goods. Although producer services do depend to a certain extent on production, the financial services sector has no direct relationship to manufacturing. In locational terms, contrasting trends have been at work. As manufacturing has dispersed worldwide, high level services have increasingly concentrated and have been doing so in places different from the old centres of manufacturing.

At the top of the urban service hierarchy are the global cities of London, New York and Tokyo. These cities are the major nuclei of global industrial and financial command functions. The economic strength of these cities has become more and more detached from the local economies in which they are located and they have become embedded in a truly global set of economic relations. The prominence of the global cities is now less as centres of corporate headquarters and more as leaders in financial markets and financial innovation. In an era of hypermobility of financial capital the volume of activity conducted in these centres is absolutely crucial to their success as margins on transactions have become increasingly slight as a result of intense competition. However, the knowledge structures and institutions that comprise the growth engines of the global cities are difficult for other cities lower down the hierarchy to capture.

Below the three global cities is a second level of about 20 cities, including Frankfurt, Paris, Brussels, Milan, Chicago and Los Angeles, which also have significant global connections. All the cities in the two top rungs of the global hierarchy offer a wide range of highly specialised services. There is a production process in these services that benefits from proximity to other specialised services. For example, the production of a financial instrument requires inputs from accounting, advertising, legal expertise, economic consulting, designers,

London: a global city

public relations and printers. Time replaces weight in this process as a force for agglomeration.

The 1990s witnessed the emergence of a growing number of transnational service conglomerates, for example Merill Lynch, seeking to extend their spatial influence as far as possible. The recent development of the global advertising industry is a case in point.

However, many lower level services are decentralising from MEDCs. A recent headline in the *Daily Telegraph* (18 February 2003) read ' Britain to lose 65,000 jobs in call centres'. These jobs were destined for India where an average of five call centres are opening each month. Staff receive training on British accents as well as advice on football teams and soap operas so that they can chat knowledgeably to callers. Pay roll costs in India are between 10% and 20% of British levels, so it is not surprising that lower and middle level services are being out-sourced to India and other lower wage economies. The globalisation of services is simply following what has been happening in manufacturing for some time. It should not be forgotten that the developed nations still dominate trade in high-level services. The major global service providers are scanning the globe for new markets to move in to, as exemplified by the title of a book published in 2001 – *China's Service Sector: A New Battlefield for International Corporations.*

6 Tourism

The package tour has achieved global extension. Waters sees the cultural impact of globalised tourism as multiple and complex with its key dimensions shown in Figure 25.

On most counts travel and tourism is the largest industry in the world. The World Travel and Tourism Council (WTTC) estimates that tourism sustains more than one in ten jobs around the world. The rapid expansion of the industry in recent decades has, however, brought the economic, environmental and social/cultural issues surrounding it to the fore. Tourism is arguably the most contentious aspect of globalisation. Evidence of the downside of tourism is such

- The extent of globalised tourism indicates the degree to which tourists themselves conceptualise the world as a single place, which is without internal geographical borders.
- Globalisation exposes tourists to cultural variation confirming the validity of local cultures and their differences.
- The objects of the tourist gaze are obliged to relativise their activities, that is, to compare and contrast them to the tastes of those that sightsee.
- Tourism extends consumer culture by redefining both human practices and the physical environment as commodities.

Figure 25 Water's key dimensions of globalised tourism

that many communities, environmental groups and human rights campaigners are very much against it. Many communities in the developing world have suffered through the imposition of the worst of western values, resulting in some or all of the following problems:

- the loss of locally owned land
- the abandonment of traditional values and practices
- displacement
- alcoholism
- prostitution, sometimes involving children
- drug abuse
- crime.

At best there is usually some degree of irritation caused by tourism, at its very worst the impact of tourism amounts to gross abuse of human rights (Figure 26). For example, the military junta in Myanmar has forcibly moved large numbers of people from their homes to make room for tourist development and used hundreds of thousands of people as forced labour on tourist-related projects.

In India the 32,000 hunter-gatherer Adivasis have had their access to the forest they lived in for centuries severely curtailed because it forms part of Nagarahole National Park. They are now not allowed to hunt or cultivate, keep livestock or collect forest produce, and visit sacred sites and burial grounds. The Adivasis, for their part, went to court to prevent the construction of a hotel in the national park. Apart from the denial of their historic rights the Adivasis are concerned about the erosion of their tribal culture.

In Hawaii, traditional burial grounds have been razed to make way for new resorts, while in Bali devout Hindus are angry that their temples are overshadowed by recently constructed large-scale tourist developments.

The creation of Kruger National Park resulted in the total denial of resources to local people. The advent of democracy in South Africa now means that this situation may change as the government looks towards a sharing arrangement that it hopes will balance local, national and international interests. The development of this process will be closely watched by a number of other developing countries.

The industry has a huge appetite for basic resources. A long-term protest against tourism in Goa highlighted the fact that one five-star hotel consumed as much water as five local villages with the average hotel resident using 28 times more electricity per day than a local person.

Tourism that does not destroy what it sets out to explore has come to be known as 'sustainable'. The term comes from the 1987 UN Report on the Environment, which advocated the kind of development that meets present needs without compromising the prospects of future generations. Following the 1992 Earth Summit in Rio de Janeiro, the WTTC and the Earth Council drew up an environmental

WHAT'S THE PROBLEM?

Far too often, displacement is the real price of tourism, and it is paid by local people. Your livelihood, traditions and culture may count for nothing if your home is on an idyllic coastline, or by a cool hill, or among ancient temples … .

If you live in a country with plenty of sunshine and a growing tourism industry, and if your land would suit a new beach complex, a golf course, a safari park or a culture and heritage site, you could be 'in the way' of profitable development. You might as well pack your bags and leave - or else you may well be forced off your own land.

> 'How could any money compare with what they have taken from us, to the grazing we have lost, to the human lives we have lost by being kept out of the Amboseli?'
>
> *(Metoe ole Lombaa Maasi Pastoralist)*

The tourism industry is big. If it was a country it would be the third richest in the world. And it's growing – five years from now it will be the world's biggest industry. There are times when it seems that absolutely nothing can stand in its way … .

DISPLACEMENT: WHAT IT REALLY MEANS TO PEOPLE

Displacement isn't a simple matter of persuasion, compulsory purchase and compensation. It's a matter of force. People just *have* to leave.

1996 – Burma's Year of the Tourist. Among Burma's (Myanmar's) many attractions is the Royal Palace at Mandalay, where the moat has been cleared – by forced 'voluntary' labour and homes demolished. And there is Pagan, where 5,200 people who lived in villages among the ancient pagodas were given two weeks to pack up and leave. Now Pagan and its pagodas welcome tourists in peace and quiet, while Pagan's people have been moved to a site of bare, parched earth with little shelter.

ISLAND PARADISE? The lovely tropical island of Lombok is increasingly popular with visitors to Indonesia. According to witnesses, the Government is tearing down homes to make way for development, and the tourism frontier is moving so fast that communities have no time to prepare and adjust … .

FAIR GAME? In Kenya, the Maasai have been forced off their ancestral lands. These places have become National Parks, where tourists are free to roam and observe big game. But the Maasai are banned. Many have migrated to urban slums. Many of those that have stayed sell souvenirs, or beg money for posing for photos.

DISPLACEMENT: WHAT TOURISM CONCERNS IS DOING

Your may be asking: "Well, what on earth can I do about it?" Just like displaced people around the world, those of us who travel with a conscience can't help but feel helpless.

But the cause isn't altogether lost. Tourism Concern is trying to make tourism fairer, to make it possible for us to travel and take holidays without costing local people their homes and livelihoods. Tourism Concern campaigns – all over the world – on behalf of people displaced by the tourism industry.

Tourism Concern *Campaigns Worldwide for Just and Sustainable Tourism • Helps Develop Projects Worldwide for Just and Sustainable Tourism • Works to Help Tourists and Travellers Understand the Issues • Forges Links with People in Travel and Tourism Destination Areas • Lobbies the Tourism Industry to Take Local People into Account.*

If you have a conscience about your travel and holidays, and if you want to help the process of making the tourism industry more responsible, more sensitive and more just, please help us with our work.

YOU CAN MAKE A DIFFERENCE

Every penny you donate helps us work for a fairer deal for the people who have to pay the greatest price for tourism.

Figure 26 The negative impact of tourism according to the pressure group Tourism Concern

checklist for tourist development, which included waste minimisation, re-use and recycling, energy efficiency and water management. The WTTC has since established a more detailed programme called 'Green Globe', designed to act as an environmental blueprint for its members.

The pressure group Tourist Concern defines sustainable tourism as 'Tourism and associated infrastructures that: operate within capacities for the regeneration and future productivity of natural resources, recognise the contribution of local people and their cultures, accept that these people must have an equitable share in the economic benefits of tourism, and are guided by the wishes of local people and communities in the destination areas'. This definition emphasises the important issues of equity and local control, which are difficult to achieve for a number of reasons:

- governments are reluctant to limit the number of tourist arrivals because of the often desperate need for foreign currency
- local people cannot compete with foreign multinationals on price and marketing
- it is difficult to force developers to consult local people?

In so many developing countries newly laid golf courses have taken land away from local communities while consuming large amounts of scarce freshwater. It has been estimated that the water required by a new golf course can supply a village of 5000 people. In both Belize and Costa Rica coral reefs have been blasted to allow for unhindered watersports. Like fishing and grazing rights, access to such common goods as beach front and scenically desirable locations does not naturally limit itself. As with overfishing and overgrazing the solution to over-touristing will often be establish ownership and charge for use. The optimists argue that because environmental goods, such as clean water and beautiful scenery, are fundamental to the tourist experience, both tourists and the industry have a vested interest in their preservation. The fact that 'ecotourism' is a rapidly growing sector of the industry supports this viewpoint, at least to a certain extent.

Education about the environment visited is clearly the key. Scuba divers in the Ras Mohammad National Park in the Red Sea, who were made to attend a lecture on the ecology of the local reefs, were found to be eight times less likely to bump into coral (the cause of two-thirds of all damage to the reef), let alone deliberately pick a piece.

A new form of ecotourism in which volunteers help in cultural and environmental conservation and research is developing. An example is the Earthwatch scientific research projects, which invite members of the general public to join the experts as fully-fledged expedition members, on a paying basis of course. Several Earthwatch projects in Australia have helped Aboriginal people to locate and document their prehistoric rock art and to preserve ancient rituals directly.

In some resorts coastal erosion has diminished beaches so much that expensive replenishment schemes have been undertaken. In 1994, a multi-million-pound scheme to replenish Skegness's dunes and beach with eight million tonnes of sand from the North Sea was begun and now the beach resembles its former fulsome state. However, the dredging has had a catastrophic effect on the shrimp catch in nearby coastal waters. Also, such replenishment is only a short-term solution because the dredging offshore eventually increases erosion on the coast, creating a vicious circle. As a result, it is predicted that the coast at Skegness will again become denuded in about 40 years time. In Morocco, sand dunes have been removed to create artificial beaches on rocky shores in the Canary Islands.

In March 1998 the issue of no-go zones for tourists was debated at the Royal Geographical Society in London. The move comes amid concern that tourism is damaging many environments and social cultures. Environmentalists and an increasing number of people in the tourist industry itself believe that tourists should be banned from some wildlife sites to prevent their degradation, and from remote areas where indigenous tribes could be affected adversely by exposure to foreigners. Other, partial bans could be introduced to curb the impact of traffic on historic towns and cities. The idea of imposing access restrictions is not, of course, completely new but the scale of restriction currently being discussed is.

Nearly all the world's most valuable coral reefs have been damaged by tourism (as well as by shipping and destructive fishing technologies). It is likely that dive tourism will be increasingly restricted

Tourism in Iceland: a fragile environment

and replaced by viewing from glass-bottomed boats instead. The Victoria Falls on the Zimbabwean–Zambian border is suffering from an increasing number of visitors undertaking a growing range of activities. This World Heritage Site has become a magnet for bungee-jumpers, whitewater rafters, sky-divers and kayakers. The rain forest in the surrounding area is suffering from trampled vegetation, soil erosion, litter and the threat of fires. The wild game is being squeezed out and local people are finding it more and more difficult to practice traditional activities. The pressure for imposing restrictions is growing.

Summary

- The Clark-Fisher model shows how, over time, the relative importance of the sectors of employment change.
- The business cycle and Kondratiev long waves help to explain short- and longer-term changes in the growth and integration of the global economy.
- Farming and food production is becoming increasingly dominated by large biotechnology companies, food brokers and huge industrial farms.
- As fish stocks have dwindled in more and more fishing areas the industrial fishing fleets of the major fishing nations have moved to more distant waters to maintain their catches, often with devastating consequences for local fishermen in LEDCs.
- The networks that carry the trade in minerals and fuels have never been more complex.
- The globalisation of services is based on a hierarchy of global cities.
- In tourism the package tour has achieved global extension.
- Tourism is arguably the most contentious aspect of globalisation.

Questions

1. **(a)** Define the terms primary, secondary, tertiary and quaternary.
   (b) Describe the changes that occur in the three stages of the Clark-Fisher model.
   (c) Explain the sharp decline in agricultural employment over time.
   (d) Why does secondary employment reach a peak in the industrial stage and then decline?
   (e) Account for the changes in the relative importance of the tertiary and quaternary sectors over time.
2. **(a)** Briefly explain the difference between the business cycle and Kondratiev long waves.

(b) Describe the technological changes associated with the development of long waves.
(c) How has the organisation of firms changed with successive long waves?
(d) Comment on the infrastructure characteristics and the geographical focus of each wave.

3. **(a)** Discuss the issues surrounding the globalisation of agriculture.
(b) Why is fishing a source of disagreement between an increasing number of countries?
(c) Explain the increasing globalisation of services.

4. **(a)** Why is tourism arguably 'the most contentious aspect of globalisation'?
(b) Discuss the ways in which tourism can be organised in a more sustainable manner.

4 Deindustrialisation and Reindustrialisation

KEY WORDS

Deindustrialisation: the long-term absolute decline of employment in manufacturing.
Reindustrialisation: the establishment of new industries in a country or region that has experienced considerable decline of traditional industries.
Economies of scale: a situation in which an increase in the scale at which a business operates will lead to a reduction in unit costs.
Product life cycle: the pattern of sales in the life of a product usually divided into four stages: early, growth, maturity and decline.
Industry mix: the industrial structure of a country or region. A positive industry mix denotes a bias towards industries that are growing at rates above the national or international average. A negative industry mix is the reverse of this situation.

1 Globalisation has caused the progressive destruction of manufacturing industry in Britain and the bulky raw materials that it needed. The parts of manufacturing that can survive are at the top quality end; the manufacturing one sees around Britain these days is mainly new and reliant
5 on innovation of one sort or another.

Patrick Minford (Professor of Economics, Cardiff Business School)

Globalisation is widely blamed for both the decline in manufacturing and the fall in demand for unskilled workers, but both trends are primarily driven by technology, not trade.

(Philippe Legrain)

1 The Declining Importance of Manufacturing in MEDCs

In the USA and Britain the proportion of workers employed in manufacturing has fallen from around 40% at the beginning of the 20th century to less than half that now. Even in Japan and Germany, where so much industry was rebuilt after 1945, manufacturing's share of total employment has dropped below 25%. Not a single developed country has bucked this trend known as deindustrialisation, the causal factors of which are:

- Technological change enabling manufacturing to become more capital intensive and more mobile.

- The filter-down of manufacturing industry from developed countries to lower wage economies, such as those of South-east Asia.
- The increasing importance of the service sector in the developed economies.

There can be little surprise in the decline of manufacturing employment for it has mirrored the previous decline in employment in agriculture in the developed world. At the beginning of the 19th century, 68% of Japan's labour force worked on farms. In the USA and Britain the figures were 40% and 20%, respectively. Today, agriculture accounts for only 7% of employment in Japan, 3% in the USA and less than 2% in Britain. So, if the decline of manufacturing in the developed world is part of an expected cycle, the consequence of technological improvement and rising affluence, why is so much concern expressed about this trend. The main reasons would appear to be:

- The traditional industries of the industrial revolution were highly concentrated, thus the impact of manufacturing decline has had severe implications in terms of unemployment and other social pathologies in a number of regions.
- The rapid pace of contraction of manufacturing has often made adjustment difficult.
- There are defence concerns if the production of some industries falls below a certain level.
- Some economists argue that over-reliance on services makes an economy unnecessarily vulnerable.
- Rather than being a smooth transition, manufacturing decline tends to concentrate during periods of economic recession.

Tate and Lyle sugar cane refinery, London Docklands

In general, high-technology manufacturing, such as computers, aerospace, pharmaceuticals and electronics, has been able to hold its share in the economies of MEDCs. However, medium- and low-technology manufacturing has declined markedly. An OECD (Organisation for Economic Co-operation and Development) survey identified six sectors where imports from low-wage economies are particularly important. These are clothing and shoes, wood products, rubber and plastics, computer equipment, vehicles other than cars and planes, and some consumer goods such as toys. These trends have been reflected clearly in the composition of trade of the richer economies.

It is difficult to say exactly how much trade has harmed unskilled workers in MEDCs. Most economists conclude that technology is responsible for much of the impact. Trade, it seems, accounts for less than a quarter of the increase in inequality between skilled and unskilled workers in MEDCs. Other factors that have also had an influence are falling unionisation, immigration and labour-market deregulation.

In the USA employment in manufacturing peaked at 22.5 million in 1979. This fell to 19.9 million in 2000. During this period the growth of employment in services was rapid and by 2000 only 14.7% of US employment was in manufacturing In the service sector employment rose from 49.0 million in 1970 to 101.9 million in 2000.

Despite all the job losses in manufacturing, unemployment was, until the economy turned down in 2001, at a 30-year low. However, very often the new jobs did not go to people who had lost traditional manufacturing jobs. Unemployment resulting from factory closures often remains persistent, creating an underclass of people who feel excluded from mainstream society.

In the UK manufacturing employment fell from 8.5 million in 1970 to five million in 1998, 18.6% of the workforce. In 1970, 39.5% of West German workers were in manufacturing; in 1999, only 24.1% of unified Germany's workforce were employed in this sector.

2 Positive and Negative Deindustrialisation

Economists have recognised two types of deindustrialisation; positive and negative. *Positive deindustrialisation* occurs when the share of employment in manufacturing falls because of rapid productivity growth but where displaced labour is absorbed into the non-manufacturing sector. In such a situation the economy is at or near full employment and GDP per capita is rising steadily. In contrast, *negative deindustrialisation* results from a decline in the share of manufacturing in total employment, owing to a slow growth or decline in demand for manufacturing output, and where displaced labour results in unemployment. Unfortunately, in the UK and many other developed countries the deindustrialisation experienced has been predominantly of the negative kind.

3 The Filter-down Process of Industrial Relocation

This process, detailed by WR Thompson and others, operates at both a global and a national scale. Economic core regions have long been vulnerable to the migration of labour-intensive manufacturing to lower wage areas of the periphery as exemplified in the USA by the historical drift of the textile and shoe industries from New England, and apparel manufacture from New York, to North and South Carolina. The filter-down process is based on the notion that corporate organisations respond to changing critical input requirements by altering the geographical location of production to minimise costs and thereby ensure competitiveness in a tightening market.

The economic core (at a national and global level) has monopolised invention and innovation, and has thus continually benefited from the rapid growth rates characteristic of the early stages of an industry's life cycle (the product life cycle), one of exploitation of a new market. Production is likely to occur where the firm's main plants and corporate headquarters are located. Figure 27, illustrating the product life cycle, indicates that in the early phase scientific-engineering skills are at a high level and external economies are the prime location factors.

In the growth phase, methods of mass production are gradually introduced and the number of firms involved in production generally expands as product information spreads. In this stage management skills are the critical human inputs. Production technology tends to stabilise in the mature phase. Capital investment remains high and the availability of unskilled and semi-skilled labour becomes a major locating factor. As the industry matures into a replacement market the production process becomes rationalised and often routine. The high wages of the innovating area, quite consistent with the high level skills required in the formative stages of the learning process, become excessive when the skill requirements decline and the industry, or a section of it, 'filters-down' to smaller, less industrially sophisticated areas where cheaper labour is available, but which can now handle the lower skills required in the manufacture of the product.

On a global scale, large transnational companies have increasingly operated in this way by moving routine operations to the developing world since the 1950s. However, the role of indigenous companies in developing countries should not be ignored. Important examples are the 'chaebols' of South Korea, such as Samsung and Hyundai, and Taiwanese firms such as Acer. Here the process of filter-down has come about by direct competition from the developing world rather than from the corporate strategy of huge North American, European and Japanese transnationals.

It has been the revolution in transport and communications that that made such substantial filter-down of manufacturing to the

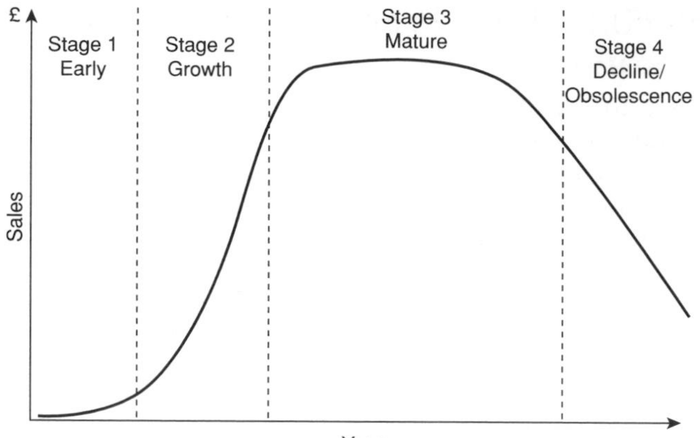

Requirements	Product Cycle Phase		
	Early	Growth	Mature
Management	2	1	3
Scientific-Engineering Know-How	1	2	3
Unskilled and Semiskilled Labour	3	2	1
External Economies	1	2	3
Capital	3	1[a]	1[a]

Figure 27 The product life cycle

developing world possible. Containerisation and the general increase in scale of shipping have cut the cost of the overseas distribution of goods substantially, while advances in telecommunications have made global management a reality. In some cases whole industries have virtually migrated, as shipbuilding did from Europe to Asia in the 1970s.

In others the most specialised work gets done in developed countries by skilled workers, the simpler tasks elsewhere in the global supply chain.

Although the theory of the product life cycle was developed in the discipline of business studies to explain how the sales of individual products evolve, it can usefully be applied at higher scales. A firm with a range of ten products, half in stage 3 and half in stage 4, would have no long-term future. A healthy multi-product firm will have a strong R&D (research and development) department ensuring a steady movement of successful products on to the market to give a positive distribution across the four stages of the model. Likewise, the industry mix of a region or a country can be plotted on the product life cycle diagram. Regions with significant socio-economic problems are invariably over-represented in stages 3 and 4. In contrast, regions with dynamic economies will have a more even spread across the model with particularly good representation in the first two stages.

4 Organisational Innovation

The manufacturing industries that have survived best in MEDCs are those which have been most innovative in the ways they operate. This has often involved taking on lean manufacturing technology. 'Lean' manufacturing techniques were first developed in the 1950s in Japan, by a Toyota manager called Taiichi Ohno (Figure 28).

Lean manufacturing seeks to combine the best of both craftwork and mass production. It seeks to use less of each input and to eliminate defects – if a fault is spotted the production line is halted

1. Strategic management: aiming to forecast and, if possible, to control the future relationship between the organisation and its supplier and customer markets.
2. Just-in-time (JIT): minimising inventory at each stage of the production process to reduce stock and storage costs.
3. Total Quality Management: ensuring that supplies of components from outside and inside the firm are reliable in quantity and high in quality.
4. Teamwork: the sharing of tasks for a small group of workers at a similar stage of the production process.
5. Managerial decentralisation: allowing quicker and more knowledgeable responses to changing market conditions.
6. A numerically flexible labour force: reducing or increasing the labour force as demand falls or rises.
7. Functionally flexible workers: involving employees in as many of the firms activities as possible.

Figure 28 The elements of Toyotism (modern Japanese industrial practices)

immediately and remedial action taken. The process eliminates waste by making only as much as is wanted at any given time. Advances in manufacturing software programs have allowed companies to integrate the various aspects of their work to a higher degree than ever before. At first JIT transferred the burden of storing parts to the suppliers of assembly plants but at its most advanced it extends all the way along the supply chain. JIT encourages suppliers to concentrate around assembly plants to ensure rapid delivery. In fact, some assembly plants insist on suppliers being no more than a certain distance away. Ideally, the number of suppliers should not be too extensive and long-term contracts should be agreed. Increasingly, production plants and suppliers are collaborating on research. As supply interruptions would prove disastrous, agreements are often signed between employers and employees to avoid such an occurrence. Single union agreements are an important part of this process.

It was not until the mid-1970s that US firms began to employ lean manufacturing, as Japanese goods were making considerable inroads into US markets. Once it was proved that such techniques could work outside Japan, large European companies followed suit in an effort to bridge the productivity gap. However, Japanese companies still lead in the productivity stakes. It is thus not surprising that Japanese car production soared from one million in 1989 to 2.5 million in 1998. Although lean manufacturing developed in the car industry, similar strategies are now employed in a wide range of industries.

The adoption of the lean system paved the way for further increases in productivity and manufacturing flexibility through the integration of IT-based AMT (Advanced Manufacturing Techniques). This replaced simple automating devices with numerically controlled tools, industrial robots and flexible-transfer machines, and eventually computer-integrated manufacturing systems. These new techniques have been largely responsible for significant advances in productivity in a number of large mature industries.

5 Reindustrialisation, Rationalisation and Restructuring

The development of new industries has, to a limited extent, offset the decline of traditional manufacturing. The sector at the forefront of reindustrialisation is high technology. In terms of the process as a whole, small firms have led the way. The increasing level of global competition has driven all industries to improve their productivity. The consequences have generally been:

- Rationalisation or 'downsizing' of the workforce with the expectation that a smaller number of workers will maintain the same level of production.
- Closure of inefficient plants.

- Restructuring by: (a) introducing more efficient production methods; (b) merging with another company in the same sector.

Large TNCs continuously compare all aspects of production in their plants in different countries and can move quickly to rationalise either when the market becomes more crowded as a result of increasing competition or during a recession when demand in general declines.

CASE STUDY: DEINDUSTRIALISATION AND REINDUSTRIALISATION IN THE GERMAN RUHR

The Ruhr coalfield region, the largest concentration of heavy industry in Europe and the place where Marshall Aid money (from the USA for European economic recovery) first ignited 'the post-war economic miracle', is in the throes of a dramatic transformation (Figure 29). Most of the coal mines, steel works and the other traditional industries on which the region's reputation was based have closed, to be replaced by dynamic modern enterprises such as telecommunications, computer products and service industries. Although this structural change has not been easy, the evidence of it can be found all across the region. Structural change, an important element of globalisation, has occurred to a greater or lesser degree in all traditional heavy industrial regions around the world but few regions have matched the transformation of the Ruhr in terms of scale and pace. Thus, it is not surprising that the Ruhr is often regarded as a model of structural change for other traditional industrial areas.

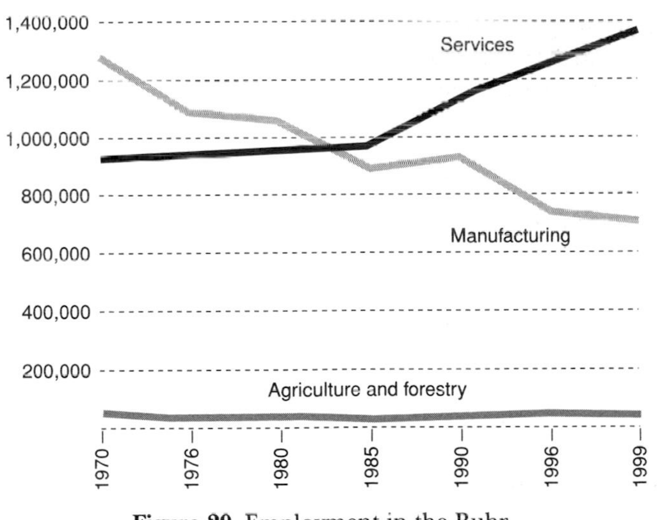

Figure 29 Employment in the Ruhr

(a) Location and Development

The Ruhr region or 'Ruhrgebiet' (4434 km²) lies along, and north of, the Ruhr river, which rises in the hills of central Germany and flows westwards to join the River Rhine at Duisburg. The Ruhr, situated in the Federal State of North-Rhine Westphalia (about 10% of its land area), is neither a historical or political entity but rather an economic and geographical area. The Association of Local Authorities in the Ruhrgebiet (KVR) is generally regarded as the statistical and geographical basis of the region. It was originally set up in 1920 as the 'Ruhr Coal Area Settlement Association'. The 11 urban areas of Bochum, Bottrop, Dortmund, Duisburg, Essen, Gelsenkirchen, Hagen, Hamm, Herne, Mulheim and Oberhausen, along with the districts of Ennepe-Ruhr, Recklinghausen, Unna and Wessel, are linked together in this single umbrella authority. From east to west the Ruhr measures 116 km and from north to south 67 km. With around 5.4 million people it is one of the largest urban conurbations in Europe. Its average population density is 1213 persons per km². The headquarters of the KVR is in Essen, which used to be the largest coal-mining city in Europe.

(b) Industrialisation

At the beginning of the 19th century the region was mainly covered in open field and woodland. The arrival of the industrial revolution changed the landscape extensively. The industrialisation of the region was based on extensive coal deposits, especially the high-quality coking coal required in steel manufacturing. Coal was found near the surface (the exposed coalfield) along the Ruhr river, where the oldest mines and steelworks were located, and at greater depths (the concealed coalfield) to the north along the Lippe river. 'Steel barons', such as Krupp and Thyssen, established their industrial empires in the Ruhr valley. During the Industrial Revolution the population grew from 274,000 in 1820 to 1.3 million in 1885 and 4.1 million in 1925, with workers coming into the area from the rest of Germany and abroad. Following a decrease in population after the Second World War the population rose to a peak of 5.7 million in 1961. Coal production reached its height in 1956 when there were almost half a million people employed in the mining industry. The crises in the coal and steel industries precipitated the subsequent fall in population, with many workers leaving for the newly developing industrial areas in south Germany. It has been projected that the region's population will decline further to 5.04 million by 2015.

(c) Deindustrialisation

In 1960, one in every three workers in the region were employed in the mining and metalworking sector. However, with (a) the loss of traditional markets to new products made from oil, synthetic materials and ceramics and (b) the filter-down of heavy industries to NICs, firms in the Ruhr began to lose orders on a large scale and unemployment increased rapidly. Since the 1970s, collieries, steelworks and other associated heavy industries have closed down one after another. The number employed in coal-mining fell from 470,000 in 1955 to 53,000 in 1999. During the same period the number of working collieries fell from 136 to nine. Coal mining is now concentrated in a few sites in the north of the region. During the 1980s and 1990s, a total of 400,000 jobs disappeared in the Ruhr. The German government tried to counter the regional crisis by providing financial aid to heavy industry, which helped to soften the impact of job losses. The region has also been in receipt of a considerable amount of EU funding through the European Regional Development Fund (ERDF) and the European Social Fund (ESF). EU funding is aimed at encouraging the development of 'small- and medium-sized enterprises, infrastructure, redeveloping disused industrial sites, improving environmental quality and the development of human resources.' The EU wants the remaining mines to close by 2005 because they are considered uneconomic and also because of their environmental impact. According to Ruth Kampherm from KVR, 'The mines are open still because ... the government of North-Rhine Westphalia holds the opinion that structural change should happen in a socially acceptable way'. However, the German government may try to negotiate an extension beyond 2005 for the most productive of the remaining mines.

Today, less than 9% of the region's employment is in the coal and steel industries. However, 31% of the coal and 11% of the steel produced in the EU still comes from the Ruhr. By 1999 the manufacturing sector's share of total employment in the Ruhr had fallen to 33.6%, down from 59.5% in 1964.

The chemical industry in the Ruhr is undergoing major changes with mergers and sell-offs prominent. Job losses are the inevitable result of such rationalisation. The 25,000 jobs that existed in the chemical industry in the Emscher-Lippe region have been reduced to just over 6000.

The merger of Degussa-Huls and the VIAG subsidiary SKW Trostberg AG in February 2001 created the world's largest special purpose chemicals business. Degussa's Marl works, one of the largest industrial sites in the Ruhr, employs more than 10,000 workers.

Even the car industry has not escaped job losses. Although the Opel works in Bochum has been working at full capacity, the company has been reducing the workforce to remain competitive. In 2000, 319,000 cars (Zafiras and Astras) were made. Opel currently employ 13,000 workers in Bochum. It is likely that this number will fall further in the future.

The rapid decline of traditional industries caused unemployment to shoot up, reaching a record of over 15% in 1987 and 1988. In 2001 it was 12.6% – considerably above the national average. However, without structural change the unemployment rate would be much higher, as much as 50% according to some commentators. Former workers in the traditional industries have been faced with three options:

- Unemployment.
- Early retirement.
- Retraining for jobs in the new lighter industries or in the service sector.

In general coal miners have found it much more difficult to secure alternative employment compared to steel workers and those in other declining industries. The long term unemployed face considerable isolation from mainstream society. Poverty is concentrated in specific urban areas and within these there are particular islands of deprivation.

(d) Reindustrialisation

Structural change can occur either spontaneously (due to market forces) or as a result of regional economic planning. In most traditional industrial areas, such as the Ruhr, spontaneous redevelopment tends to be rather limited and very spatially selective in nature, and as a result a regional economic planning approach is essential if widespread reindustrialisation is to occur. The KVR undertakes this vital role in the Ruhr. It develops and promotes regional initiatives and undertakes essential groundwork, from collating information and preparing analyses to the development and implementation of projects.

In the Ruhr, work on creating a new economic structure began in the 1960s:

- New universities and polytechnics were set up.
- The transport infrastructure was significantly improved.
- New business sectors such as chemicals, power, motor vehicles, machine building and environmental technology began to replace coal and steel.
- Information and telecommunication businesses increasingly opted to locate in the region.
- Many small- and medium-sized businesses moved into the Ruhr, operating in a more innovative manner than traditional large-scale firms. These companies operate in fields such as precision mechanics and optics, legal and business consulting, publishing, construction, and hotels and catering.

- Environmental technology has become a strong element of the Ruhr's economy. The environmental protection industry began in the region in the 1960s.
- The renewable energy industry is making its mark in the Ruhr. Gelsenkirchen boasts the world's most up-to-date solar cell factory, which commenced production in 1999.
- Medical technology (e.g. developing intelligent technologies for the early diagnosis of sickness), based mainly in the Ruhr Technology Centre in Bochum, is another expanding field. However, this industry is also represented in Duisburg, Essen, Hamm, Schwerte and Castrop-Rauxel.
- Dortmund has become an important centre for young software development firms. The town is planning to create up to 70,000 new jobs by 2010 in information technology, e-commerce and microscopy.
- Duisburg is a significant logistics centre because of its good transport connections, particularly its port with connections to the North Sea. Duisburg boasts the largest inland harbour in the world.
- The Deutsche Bergbautechnik (German Mining Technology), based in Lunen, tailors German mining products to the export market, taking account of the different conditions in each of the countries it trades with. Collieries in Chile, for example, have very different technical requirements to mines in South Africa. The knowledge gained in over a century of mining in the Ruhr is in demand in a variety of foreign markets.

Today 65% of the Ruhr's workforce are employed in the tertiary sector. However, this growth has not been enough to compensate for the loss of manufacturing jobs. The strongest branch in the service sector is the retail industry. This is followed by health and veterinary services, transport and communications, legal and business consultants, and estate agencies. Further growth areas are computer and internet services, multimedia, advertising agencies, telecommunications and engineering consultants.

Many businesses in the 'new economy' are clustered on modern trading estates or office parks. Unlike conventional industrial estates the more recent business clusters are generously laid-out with innovative architecture and landscape design. However, not all are on greenfield locations. Many empty industrial and colliery sites have been prepared for re-use by the State Development Company of North-Rhine Westphalia. The KVR's Ruhr Site Data Bank provides information on the availability of trading and industrial estates in the region.

Reindustrialisation has not occurred evenly in the region. The southern part of the Ruhr has been much more successful than the north in this process. More plant closures have been concentrated in the north, which also suffers from a poorer infrastructure than the south.

(e) Industrial Reclamation

The closure of coal mines, steel works and other heavy industrial premises resulted in large areas of abandoned industrial wasteland. These areas contain a great variety of different types of subsurface soil, most of it man-made and chemically polluted. However, the scale of industrial reclamation in this region has been staggering. For example, an entire region along the River Emscher was transformed within the framework of the 'Internationale Bauasstellung' (IBA), an organisation responsible for the reclaiming and renewal of industrial wasteland. Here, 110 projects are spread over an area of 800 km^2.

Among the more spectacular visual examples of renewal are:

- In Duisburg a disused iron and steel works is now a gigantic 200-hectare industrial museum.
- In Oberhausen a 117-metre high gasometer has been transformed into the highest exhibition hall in Europe.
- In Mulheim, the water tower Aquarius which once held 500,000 litres of water is home to a multimedia water theme museum.
- The first large shopping centre in Germany, the Rhine–Ruhr Zentrum, was developed on the former site of a coal mine in Mulheim.

The Ruhr is attracting a growing number of tourists. The Route of Industrial Heritage, opened in 1999, shows 150 years of industrial history over its 400 km length. Structural change has provided the opportunity for the expansion of nature conservation in the region, which now boasts 300 conservation areas covering more than 4% of the total land area.

The NRW State Environment Office monitors atmospheric pollution in the region. Deindustrialisation has had a positive impact on air quality. For example, the average level of sulphur dioxide emissions fell from 66 $\mu g/m^3$ in 1981 to 8 $\mu g/m^3$ in 1999. The amount of airborne particles almost halved during the same period.

(f) The Ruhr and 'The German Problem'.

Reindustrialisation in the Ruhr must be seen in the context of the significant difficulties the German economy has faced over the last decade. For long the leading economic light of Europe, Germany is now 'the sick man of Europe', according to Otmar Issing, the European Central Bank's chief economist. In mid-2002 unemployment stood at four million or 9% of the workforce, with Germany trailing the rest of Europe in economic growth and job creation. Some economists fear that Germany will slide into a Japanese-like spiral of deflation and economic stagnation. In the 2002 World Competitiveness Report published by the International Institute of Management Development, Germany ranked 47 out of 49 industrialised countries in 'adaptability and flexibility', characteristics that are vital for the successful attraction of foreign direct investment. A recent article on the German economy suggested that it needed to do the following:

- Create more incentives for small businesses, the prime source of new jobs.
- Rewrite laws that hold back entrepreneurs.
- Loosen labour laws to attract a higher level of investment.
- Revamp and simplify the country's cumbersome tax code.
- Deregulate to encourage competition.

As the process of globalisation proceeds, Germany and the Ruhr will have to continue to adapt to changing economic conditions if they are to maintain the prosperity of the past.

Summary

- Manufacturing employment in MEDCs has fallen substantially in recent decades due to technological change, the filter-down of industries to NICs and the increasing importance of the service sector.
- Deindustrialisation has caused significant economic and social problems in many traditional industrial regions.
- The greatest job losses have occurred in labour-intensive, low-technology industries.
- Manufacturing employment peaked in Britain in 1966 and in the USA in 1979.
- Two types of deindustrialisation, positive and negative, can be identified.
- The concepts of the product life cycle and filter-down help to explain the 'global shift' of manufacturing industry.
- The manufacturing that has survived best in MEDCs has been innovative in the ways in which it operates.
- Some regions have been relatively successful in reindustrialisation while others have struggled to attract enough new industry.
- The Ruhr is the largest industrial region in Europe.
- Although structural change has been far-reaching in the Ruhr, unemployment in the region remains considerably above the national average.

Questions

1. **(a)** Define 'deindustrialisation'.
   (b) What is the evidence that deindustrialisation has occurred in MEDCs?
   (c) Briefly discuss the causes of deindustrialisation.
   (d) Why do the consequences of deindustrialisation vary from region to region?
2. **(a)** Define the 'product life cycle'.
   (b) Describe the change in volume of sales with each stage of the product life cycle.
   (c) Discuss the changing relative importance of production factors in the product life cycle.
   (d) How might such changes influence industrial location?
3. **(a)** Why did the Ruhr become such an important industrial area?
   (b) Describe the deindustrialisation of the Ruhr.
   (c) How successful has reindustrialisation been in the region?
   (d) Discuss the environmental impact of deindustrialisation and reindustrialisation.

5 Transnational Corporations

KEY TERMS

Transnational Corporation: 'a firm that has the power to co-ordinate and control operations in more than one country, even if it does not own them' (Peter Dicken).
Foreign direct investment: overseas investments in physical capital by transnational corporations.

1 Global Expansion

Transnational corporations are the driving force behind economic globalisation. They are capitalist enterprises that engage in foreign direct investment and organise the production of goods and services in more than one country. As the rules regulating the movement of goods and investment have been relaxed, TNCs have extended their global reach. As the growth of foreign direct investment has expanded, the sources and destinations of that investment have become more and more diverse. There are now few parts of the world where the direct or indirect influence of TNCs is not important. In some countries and regions their influence on the economy is huge. Apart from their direct ownership of productive activities, many TNCs are involved in a web of collaborative relationships with other companies across the globe (Figure 30). Such relationships have become more and more important as competition has become increasingly global in its extent.

Nike does not make any clothes or shoes itself. It contracts out production to South Korean and Taiwanese companies. These companies operate not only in their home countries but also in lower wage Asian economies such as the Philippines and Vietnam. Nike's expertise is in design and marketing. In 2001 Nike's total sales were $9.5 billion with profits of $590 million (6.2% of sales).

The figures supplied by Nike for its cost/price chain are as follows:

- Contractors are paid an average of $18 a shoe by Nike. This is made up of $11 for materials, $2 for labour, $4 for other costs, and $1 for profit.
- Nike sells the shoes to retailers for $36. The mark up of 100% accounts for the costs of design, research and development, marketing, advertising, shipping, production management, other sales and business costs, taxes and of course a profit.
- Retailers mark to another 100% to $72 (on average) to cover wages, shrinkage, insurance, advertising, supplies and services, depreciation, taxes and profit.

Figure 30 Nike: out-sourcing production

Company	Revenue ($M)
1. Wal-Mart Stores	219,812
2. Exxon Mobil	191,581
3. General Motors	177,260
4. BP	174,218
5. Ford Motor	162,412
6. Enron	138,412
7. Daimler Chrysler	136,897
8. Royal Dutch/Shell Group	135,211
9. General Electric	125,913
10. Toyota Motor	120,814
11. Citigroup	112,022
12. Mitsubishi	105,814

Figure 31 The world's largest corporations 2001 (Source: *Fortune*)

TNCs have a substantial influence on the global economy in general and in the countries in which they choose to locate in particular. They play a major role in world trade in terms of what and where they buy and sell. A not inconsiderable proportion of world trade is intra-firm, taking place within TNCs. The organisation of the car giants exemplifies intra-firm trade with engines, gearboxes and other key components produced in one country and exported for assembly elsewhere.

According to UNCTAD (United Nations Conference on Trade and Development) there are now over 40,000 corporations in the world whose activities cross national borders; these firms ply overseas markets through some 250,000 foreign affiliates. The intense global competition for market share has lead to the growing domination of the largest companies. The ten largest corporations in their field now control:

- 86% of the telecommunications sector
- 85% of the pesticides industry
- 70% of the computer industry
- 35% of the pharmaceuticals industry.

Although the first companies to produce outside their home nation did not emerge until the latter half of the 19th century, by 1914 US, British and mainland European firms were involved in substantial overseas manufacturing production. Prior to the First World War the UK was the major source of overseas investment, the pattern of which was firmly based on its empire. Between the wars TNC manufacturing investment, particularly American, increased substantially (Figure 32). By 1939 the USA had become the main source of foreign investment in manufacturing. The USA was to become even more powerful in the global economy after the Second World War, for it was the only industrial power to emerge from the conflict stronger rather than weaker.

However, the USA does not dominate the global economy today in the way it did in the immediate post-war period. The reconstruction

The offices of TNCs lining São Paulo's Avenida Paulista

of the Japanese and German economies resulted in both countries playing a significant transnational role by the 1970s, which was to expand considerably in the following decades. In fact, the large Japanese TNCs were to become models for their international competitors as they revolutionised business organisation. Other developed countries such as Britain, France, Italy, the Netherlands, Switzerland, Sweden and Canada also played significant roles in the geographical spread of foreign direct investment. More latterly NICs,

Period	Type	Characteristics
1500–1800	Mercantile capitalism and colonialism	Government backed chartered companies.
1800–75	Entrepreneurial and financial capitalism	Early development of supplier and consumer markets. Infrastructural investment by finance houses.
1875–1945	International capitalism	Rapid growth of market-seeking and resource-based investments.
1945–60	Transnational capitalism	FDI dominated by USA. TNCs expand in size.
1960–present	Globalising capitalism	Expansion of European and Japanese FDI. Growth of inter-firm alliances, joint ventures and outsourcing.

Figure 32 Stages in the evolution of TNCs

such as South Korea and Taiwan, have expanded their corporate reach, not just to lower wage economies but also into MEDCs. The global economy today is undoubtedly multipolar.

Citigroup (Figure 31) is the most profitable bank in the world with operations in more than 100 countries. The bank is constantly reviewing its global network, expanding in countries where potential profits are greatest and sometimes scaling back in nations where the outlook may not look so rosy. However, the company has always been extremely careful about not leaving countries as it can be a difficult process to get a banking license back from a host government. In November 2002, Citigroup became the first US bank to open a retail branch in Russia with services including internet and telephone banking. Shortly after this the company signed a 50–50 joint venture with the Shanghai Pudong Development Bank to market credit cards. The logic of the joint venture is that it allows Citigroup to begin the process about four years before the company could have done it on their own. This strategy has in fact allowed it to jump the competitive queue in the credit card business.

2 Foreign Direct Investment

The share of developing countries in the global stock of inward direct investment is 32% (Figure 33). However, this is very unevenly distributed: 19% is in Asia but just 2.3% in Africa. According to Martin Wolf (*Financial Times,* 25 September 2002), 'Capital flows from rich countries to other rich ones or to the already successful among the poor. As a result a large proportion of humanity find itself locked inside bad locations with poor policies and worse governance. Big changes are needed if the opportunities afforded by economic integration are to become real for a larger proportion of the world's population. The starting point has to be with turning those vicious spirals of gover-

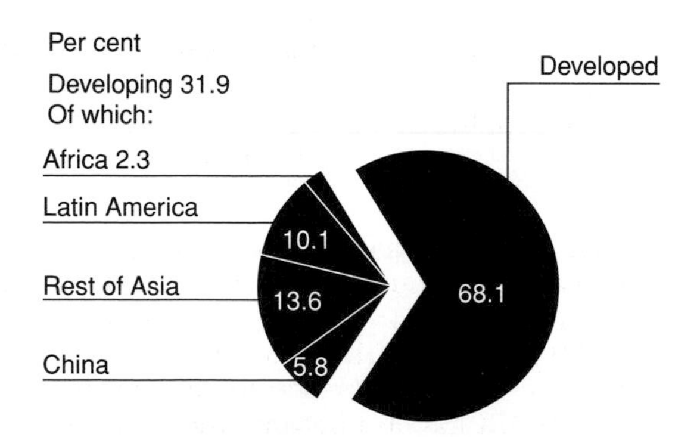

Figure 33 Accumulated total of inward direct investment flows by region, 2001. Source: *Financial Times,* 25 September 2002

nance and economic performance into virtuous ones. On the world's ability to achieve this largely depend hopes for a wider spread of successful development'.

In 2000, purchasing power parity investment per head:

- in high income countries, with 900 million people, was just over $6000
- in middle-income countries, with 2.7 billion people, was $1350
- in low-income countries, with 2.5 billion people, was under $400.

The phenomenal growth in foreign direct investment (FDI) is the most obvious sign of the increasing integration of the world's economies and much of this investment is by transnational corporations. Most FDI is in the developed world but investment in developing economies has also risen substantially in the 1980s and 1990s. Of the emerging economies, the big five in order of importance, all with an FDI stock of over $50 billion, are China, Brazil, Mexico, Singapore and Indonesia. The UN expects the growth in FDI to continue as more governments are liberalising their investment rules to attract FDI in the quest for capital and growth.

In 1997 the stock of FDI in Britain stood at $345 billion, the highest for any economy outside the USA. The inflow has been high in recent years, reaching £16 billion in 1996 and taking the total for the five-year period, 1992–6, to more than £53 billion. The figure for 1997 was even higher at £23 billion. There are now in excess of 2400 foreign-owned manufacturing firms in the UK, representing 26% of net output and 17% of the manufacturing workforce. The attractions of a UK location as cited by foreign companies are:

- relatively low levels of corporate and personal taxation
- labour flexibility
- access to the EU market
- a less oppressive regulatory climate than in many other countries.
- a stable economy with low inflation
- welcoming national, regional and local agencies involved in economic regeneration
- an attractive quality of life
- the English language, the second language for so many foreign executives.

The nature of FDI in Britain has changed over the years. Initially, branch plants carrying out routine tasks dominated foreign investment but increasingly research and development has located here. By 1997 there were 150 Japanese R&D centres in Britain. Microsoft, the computer giant, has chosen Cambridge for its first high-tech plant outside the USA. The UK has also become the favoured location for international companies to site their pan-European call centres and headquarters.

Japanese FDI in Britain: Suzuki in Crawley

However, by the latter part of 1998 evidence was growing that Britain was probably no longer the magnet for foreign investment that it once was with a number of foreign companies announcing plant closures or abandoning plans for expansion.

Britain is the world's second largest outward investor with a total outflow of almost £98 billion in the period 1992–6. Such outward investment strengthens the global reach of UK companies, bringing in substantial flows of money from profits made overseas.

Any serious analysis of FDI has to look at its targets. It is the quality of FDI rather than its quantity that brings net benefits to the receiving country. To bring such benefits FDI needs to be channelled into productive rather than speculative activities. The power of governments to influence the quality of investment has been steadily declining. It is a fact that a significant proportion of FDI is made up of companies:

- buying out state firms
- purchasing equity in local companies
- financing mergers or acquisitions.

Over the last decade or so a relatively new phenomenon has gathered pace – alliances of capital. This involves a great variety of negotiated arrangements: cross-licensing of technology among corporations from different countries, joint ventures, secondary sourcing, off-shore production of components and cross-cutting equity ownership.

3 From National to Transnational

Large companies often reach the stage when they want to produce outside of their home country and take the decision to become transnational. The benefits of such a move include:

- cheaper labour, particularly in developing countries
- exploiting new resource locations
- circumventing trade barriers
- tapping market potential in other world regions
- avoidance of strict domestic environmental regulations
- exchange rate advantages.

A particular concern of TNCs is the uncertainty of the level of future production costs in different locations. A strategy to reduce such potential risk is to locate similar plants in a variety of locations and then adopt a flexible system of production allocation between plants.

Many economists hold the view that at the global scale labour is the single most important location-specific factor. This significance is reflected in a number of ways:

- geographical differences in wage costs
- geographical variations in labour productivity (unit costs)
- geographical contrasts in the extent of labour controllability.

The first specific theory to explain why companies succeeded in the change from being wholly national entities to transnationals was produced by Stephen Hymer in 1960. The essentials of his argument were:

- A firm wishing to produce successfully in a foreign market would have to possess firm-specific advantages, which would offset the advantages held by indigenous firms. The latter should have a better understanding of the local business environment – the nature of the market, business customs, national and local legislation, etc.
- The firm-specific advantages of large international companies would focus on economies of scale, marketing skills, technological expertise and access to lower cost sources of capital.
- Such advantages would enable transnational companies to outcompete domestic firms in their own national markets.

4 Branding and the Mass Media

The spread of a global consumer culture has been important to the success of many TNCs. The mass media has been used very effectively to encourage consumers to 'want' more than they 'need'. In the modern age of sophisticated advertising, minute differences between products or small improvements in them can determine variations in demand. The power of brands and their global marketing strategies cannot be underestimated. This is particularly so in food, beverages and fashion. The increasing knowledge of Western consumer culture in the former Soviet Union and eastern Europe in the Cold War period was an important factor in the eventual disintegration of the Eastern Bloc.

5 The Importance of Place

Peter Dicken argues that far from being 'placeless' organisations as is often claimed, TNCs continue to reflect many of the basic characteristics of their home country environment. He asserts that TNCs 'are produced through an intricate process of embedding in which the cognitive, cultural, social, political and economic characteristics of the national home base play a dominant part'. However, this is not to say that companies do not take on some of the characteristics of other countries they locate in. To a certain extent it would be very difficult not to do so, but also it may be good marketing strategy to adopt certain characteristics of a host nation.

6 Variable Characteristics

The very large global corporations in particular have three basic characteristics. First, they co-ordinate and control the stages of production within and between different countries. Second, they have the ability to take advantage of geographical differences in a wide range of cost factors. Third, they have the ability to relocate resources and operations in a relatively short space of time.

TNCs vary widely in overall size and international scope. According to Dunning (1993) their transnational characteristics vary with regard to:

- the number of countries
- the number of subsidiaries
- the share of production accounted for by foreign activities
- the degree to which ownership and management are internationalised
- the division of research activities and routine tasks by country
- the balance of advantages and disadvantages to the countries in which they operate.

Dunning has identified a number of 'true global industries' that are dominated by large corporations active in all of the world's largest economies. In order of important these are: petroleum, cars, consumer electronics, tyres, pharmaceuticals, tobacco, soft drinks, fast food, financial consultancies and luxury hotels. In addition to this list are the emerging multinational alliances in airlines, telecommunications, and banking and insurance. These involve lower levels of FDI but high levels of managerial co-ordination.

7 Organisational Levels

Large TNCs often exhibit three organisational levels – headquarters, research and development, and branch plants. The headquarters of a TNC will generally be in the developed world city where the company was established. Research and development will most likely be located here too or in other areas within this country. It is the branch plants that are the first to be located overseas. However, some of the largest and most successful TNCs have divided their industrial empires into world regions, each with research and development facilities and a high level of decision making.

A recent OECD report highlighted the growing internationalisation of production through global supply chains. TNCs base different parts of their operations in different countries to reduce costs as much as possible. Foreign direct investment has increased the importance of international trade within firms. One-third of both the USA's and Japan's total trade takes place within TNCs and their affiliates. For many TNCs foreign direct investment has also become a substitute for trade. For example, a Japanese car-maker will produce locally in America rather than exporting there. European firms' dollar sales from their American subsidiaries are now four times bigger than their exports to America. Thus, a downturn in the American economy will affect Europe more than in the past.

8 Comparing the Economic Power of TNCs and Countries

The calculations of the relative size of corporations (Figure 34) used by many critics of globalisation come from the left-of-centre Institute for Policy Studies in Washington, DC. Martin Wolf (*Financial Times*, 6 February 2002) states that these calculations 'rest on an elementary howler' with the might of corporations being judged by sales but that of countries by value-added (GDP). According to Wolf the incompatibility of these two measures, grossly overestimating the power of corporations, has led to a 'paranoid delusion'.

All TNCs have to operate within national and international regulatory systems. According to Peter Dicken the 'new geo-economy' is being structured and restructured by the complex, dynamic interactions between companies and governments. However, many writers see a growing imbalance between national governments and TNCs in favour of the latter: 'As corporations gain the upper hand, the fear of job losses and the resulting social devastation has created a downward pressure

Corporations are much more than purveyors of the products we all want; they are also the most powerful political forces of our time ... Shell and Wal-Mart bask in budgets bigger than the gross domestic product of most nations ... of the top hundred economies, fifty-one are multinationals and only forty-nine are countries.

Naomi Klein, 'No Logo'

It is simply not true. The statement (opposite) is arrived at by comparing companies' sales and countries gross domestic product (GDP). But this is like comparing apples and pears. A less misleading comparison – between companies' value-added and countries value-added, their GDP – reveals that only two companies make it into the top fifty creators of value-added, and thirty-seven into the top hundred.

Philippe Legrain, 'Open World: The Truth About Globalisation'

Figure 34 Countries and corporations: alternative views

on environmental standards and social programs – what critics of unregulated corporate power call 'a race to the bottom' (W Elwood).

Daniel Litvin argues that despite the global appearance of power and competence, transnationals usually end up out of their depth. Using examples from the 16th century to the present day he shows that these large corporate entities were much less in control of events than either they or their critics would wish to believe.

Although nation states have become less powerful in some ways they remain of fundamental importance politically. They also exercise a range of constraints over companies through competition law and other regulations. Taxation is an important part of the picture. Corporate taxes make up a bigger share of government revenue today than 20 years ago.

Figure 35 summarises the possible positive and negative effects of TNCs locating in developing countries. Although TNCs are frequently criticised for exploiting labour across the globe, workers in foreign firms in poor countries earn twice the national average.

A major criticism of TNCs is that they may undermine the push for democracy in dictatorial nations in their desire for political stability. As Benjamin Barber says in McWorld and Jihad 'Capitalists may be democrats but capitalism does not need or entail democracy'.

9 Research and Development

Of the 600 companies with the largest R&D spending in the world, 275 are American, 132 Japanese, 40 German and 36 British. A DTI report published in 2002 noted that UK companies are investing a lower share of their revenues in R&D than their rivals in the USA, Japan, France, Germany and Switzerland. In the UK, R&D spending as a percentage of sales is just 2.5%. The average for the five countries

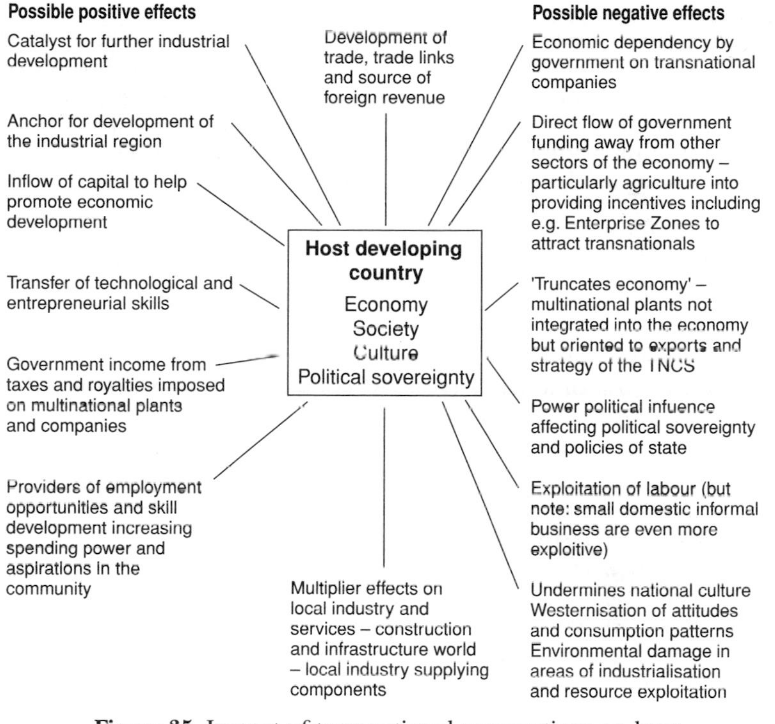

Figure 35 Impact of transnational corporations on host developing countries

named above is 5.1%. Only UK pharmaceutical companies and aerospace and defence companies are investing at above the international average. Figure 36 shows the top 12 research and development companies in both the world and the UK in 2002.

UK 2002 (1992)	**WORLD 2002** (1992)
1. **GlaxoSmithKline** (ICI)	1. **Ford** (General Motors)
2. **AstraZeneca** (Glaxo)	2. **General Motors** (Daimler Benz)
3. **BAE Systems** (Shell)	3. **Siemens** (Siemens)
4. **Unilever** (GEC)	4. **Daimler Chrysler** (IBM)
5. **Marconi** (SmithKline Beecham)	5. **Pfizer** (Hitachi)
6. **BT** (Unilever)	6. **IBM** (Ford)
7. **Rolls-Royce** (BP)	7. **Ericsson** (Toyota)
9. **Reuters** (BT)	9. **Motorola** (Matsushita Electric)
10. **BP** (Wellcome)	10. **Matsushita Electric** (Fujitsu)
11. **Invensys** (R-Royce)	11. **Cisco** (Toshiba)
12. **Amersham** (Lucas)	12. **GlaxoSmithKline** (Philips)

Figure 36 The top 12 research and development companies

10 The Growing Importance of Logistics

Logistics is the management of the flow of materials through an organisation, from raw materials to finished goods. As companies are seeking to eliminate both incoming and outgoing inventory they are outsourcing more of what they do to logistics companies.

Inventories are being reduced by:

- outsourcing more production
- buying in sub-assemblies rather than individual parts
- trying to build only to orders from customers, known as BTO (Built to Order).

Organising the supply of incoming parts and outgoing goods can account for 10% of a firm's costs. As companies have stretched their operations further and further around the world the importance of global logistics has become more and more crucial.

11 TNC Influence on United Nations Organisations

There is growing concern in many quarters that UN agencies, which have to listen to opinion from all sides, are wide open to manipulation and infiltration (Figure 37). A recent report by the World Health Organisation (WHO) stated that junk food and fizzy drinks are making children obese, and that governments should clamp down on them. The rising tide of obesity is killing and impacting severely on the quality of life of millions in the rich world and is now edging into poor countries to co-exist with malnutrition. The large TNCs in the food and drink industry do not accept this claim and are using all the influence they can muster to challenge the conclusions of the WHO.

In 2001 sales of soft drinks in the UK alone totalled £8.6 billion. A total of 21% of seven to ten-year olds drink nearly ten cans of fizzy drink a week yet one can of drink may contain up to 11 teaspoons of sugar. In the UK 9% of boys and 13.5% of girls are overweight. For every hour that they watch TV children are exposed to an average of ten food commercials. In the USA more than $54 billion was spent on 14 billion gallons of soft drinks in 1997.

The WHO report recommends that:

- Governments should clamp down on TV ads pushing 'sugar-rich items to impressionable children'.
- Governments should consider heavier taxes on such products.
- Schools should scrap vending machines.

WHO 'infiltrated by food industry'

The food industry has infiltrated the World Health Organisation, just as the tobacco industry did, and succeeded in exerting "undue influence" over policies intended to safeguard public health by limiting the amount of fat, sugar and salt we consume, according to a confidential report obtained by the Guardian.

The report, by an independent consultant to the WHO, finds that:

- food companies attempted to place scientists favourable to their views on WHO and Food and Agricultural Organisation (FAO) committees
- they financially supported non-governmental organisations which were invited to formal discussions on key issues with the UN agencies
- they financed research and policy groups that supported their views
- they financed individuals who would promote "anti-regulation ideology" to the public, for instance in newspaper articles.

"The easy movement of experts – toxicologists in particular – between private firms, universities, tobacco and food industries and international agencies creates the conditions for conflict of interest," says the report by Norbert Hirschhorn, a Connecticut-based public health academic who searched archives set up during litigation in the US for references to food companies owned or linked to the tobacco industry.

He finds that there is reasonable suspicion that undue influence was exerted "on specific WHO/FAO food policies dealing with dietary guidelines, pesticide use, additives, trans-fatty acids and sugar.

"The food industry is considerably engaged in genetically modified foods and the tobacco industry has studied the matter closely with respect to its product; there is evidence the tobacco industry planned also to influence the debate over biotechnology."

The WHO and FAO need the scientific input of the food industry, says the report, but that input must be transparent and subject to open debate.

"One industry-led organisation, International Life Sciences Institute (ILSI), has positioned its experts and expertise across the whole spectrum of food and tobacco policies: at conferences, on FAO/WHO food policy committees and within WHO, and with monographs, journals and technical briefs."

Some of the strongest criticism in the report is levelled against the ILSI, founded in Washington in 1978 by the Heinz Foundation, Coca Cola, Pepsi Cola, General Foods, Kraft (owned by Philip Morris) and Procter & Gamble. Until 1991 it was led by Alex Malaspina, vice-president of Coca Cola.

Dr Malaspina established ILSI as a non-governmental organisation "in official relations" with the WHO and secured it 'specialised consultative status" with the FAO.

Eileen Kennedy, global executive director of ILSI, said that the funding of its regional groups came exclusively from industry, while the central body received money from the branches, from government and from an endowment set up by Dr Malaspina. Nonetheless, she said, ILSI regarded itself as an independent body.

Figure 37 Influence on United Nations organisations. Source: *The Guardian*, 9 January 2003

The British Soft Drinks Association denies the link between its products and obesity, attributing the latter to sedentary lifestyles;

> Perhaps more likely than anything else to concentrate food industry minds is the spectre in the US of the courts. Two teenage boys have just filed a suit blaming McDonald's for their obesity (*The Guardian*, 2, 9 January 2003)

Previously the WHO has accused the tobacco TNCs of sabotaging efforts to control tobacco consumption through pressure tactics against the agency and other international organisations. A WHO report accused the tobacco industry of:

- Using numerous third-party organisations such as trade unions to try to influence the WHO.
- Secretly funding 'independent' experts to conduct research and publish papers that would challenge WHO findings.
- Setting up press conferences to draw attention away from events organised by the WHO related to anti-smoking efforts.

CASE STUDY: WIMBLEDON TENNIS BALLS: A TRANSNATIONAL PRODUCT

The Dunlop Slazenger balls used at Wimbledon (48,000) and many other major tennis tournaments are the product of materials and labour from at least ten different countries (Figure 38). Slazenger provided its first hand-sewn, wool-coated balls to Wimbledon in 1902. Today, tennis balls account for around a

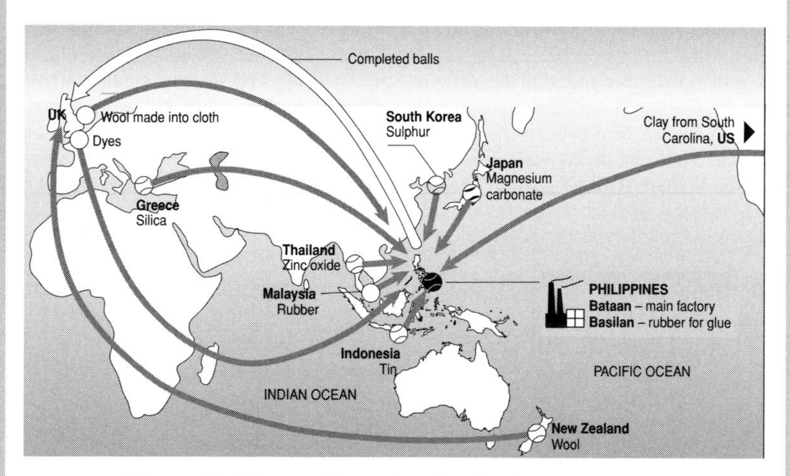

Figure 38 The making of a Wimbledon tennis ball

fifth of the £165 million annual turnover of Dunlop Slazenger who produce about 20% of the 240 million tennis balls manufactured worldwide each year.

The Philippines is now the focal point of production. The Dunlop Slazenger factory is at Bataan on the island of Basilan in the southern Philippines. Apart from manufacturing there is also a laboratory dedicated to the development and testing of tennis balls at Bataan. The engineering department at the University of Loughborough and professional tennis managers in Cheltenham are also involved in the development and testing process. In the 1970s the Philippine government set up the Bataan Economic Zone in order to attract foreign investment. However, the economic zone never attracted the amount of investment initially hoped for and today much of it is derelict as companies have left for locations such as China where labour is even cheaper. The main materials used at the Bataan plant are:

- The rubber in the core of the ball comes from Malaysia, where it is delivered to a processing plant in Prai, near Penang, from small rubber plantations in the area. The core of a ball is 40% rubber and 60% filler and chemicals.
- In addition, rubber from Basilan is mixed with petroleum naphthalene to make glue for the balls.
- The greenish yellow cloth covering (wool and synthetic mix) is made in the UK at Dursley in Gloucestershire.
- The wool for the cloth is imported into the UK from New Zealand.
- The tins into which the balls are packed are from Indonesia.

Among the substances used to vulcanise the rubber and give it the right amount of stretch and bounce are:

- clay from South Carolina
- sulphur from Korea
- silica from Greece
- magnesium carbonate from Japan
- zinc oxide from Thailand.

From the 1940s until 2002 the company produced tennis balls in Barnsley, Yorkshire. However, all production was transferred to the Philippines in 2002 to reduce labour costs, yet another example of deindustrialisation and the filter-down of production from an MEDC to an LEDC.

CASE STUDY: McDONALD'S TRANSNATIONAL FRANCHISING

With 31,000 restaurants in 121 countries, the McDonald's hamburger chain has for long been the epitome of American multinational success, as well as a symbol of US imperialism for radical groups in many parts of the world. In spite of some recent, well-publicised problems McDonald's is still the king of fast food, serving around 46 million customers each day. The company logo, the set of Golden Arches, is one of the best-known business symbols in the world. McDonald's current annual systemwide revenues total around $40 billion. When the first McDonald's restaurant was opened in Britain in 1974, people queued for hours to get in. In 1994, the drive-through line on opening day in Kuwait City was seven miles long. The busiest McDonald's outlet is in Moscow's Pushkin Square. When it opened in January 1990 it set the company record, which it still holds for the largest number of servings in a single day. It has been estimated that:

- one in eight Americans has worked in a McDonald's at some point in their lives
- a third of all cows reared in the USA are needed to produce the company's burgers
- 8% of the US potato crop is used for the fries.

Initially the ingredients for the hamburgers were prepared in individual restaurants. However, from the late 1960s the meat, bread and fries have been mass produced and then delivered frozen to each franchise outlet.

McDonald's in Moscow

The company has sought rapid growth since its inception in 1955, when Ray Kroc first franchised a company that had originated in 1937 as a California burger stand run by the McDonald brothers. Between 1965 and 1991 average annual revenue growth of 24% was based on innovative marketing and the enforcing of rigid standards of quality and cleanliness. Marketing was based on what one expert has described as the 'aspirational thing' – if you could eat hamburgers and drink Coke, you could taste part of the American dream.

However, as competition within the USA grew stiffer, McDonald's increasingly turned overseas to expand. In 1991, for the first time, McDonald's opened more outlets outside the USA than inside. In 1996, the peak year of expansion, the company opened 2000 restaurants globally. But the picture has not been so rosy in recent years. The US market is virtually saturated and the company seems to have expanded too fast in countries where not enough people can afford a $1 hamburger. David Upton, the Harvard business professor and author of an MBA thesis on the company says 'McDonald's is suddenly reaching the boundaries of growth'.

In late 2002 McDonald's announced to the Wall Street stock market that it was about to post its first quarterly loss in 37 years as a public company (Figure 39). After closing 163 restaurants in 2001 it has recently announced the closure of 175 outlets including 35 in Turkey alone. However, elsewhere other new restaurants will continue to open. In 2003 the company planned to add 300 new outlets to the 6000 already in place in Europe. There are 500 outlets in China, the focus of much recent expansion, with more than 100 estimated to open soon.

The spatial strength of McDonald's is not, however, as strong as most people think. Despite having the majority of its outlets overseas, McDonalds earns between 55% and 60% of its operating income in the USA. A total of 80% of its sales are generated

McDonald's makes a loss for first time

McDONALD'S, the fast-food chain, yesterday reported is first ever quarterly loss.

The company, which has more than 1,200 restaurants in Britain alone, had a pre-tax deficit of £198.6 million and an operating loss of of £125.6 million in the last three months of 2002.

It made a net loss of £212 million over the period but still managed a turnover of £2.4 billion.

Friends of the Earth said the loss indicated a change in attitude towards McDonald's.

Jim Cantalupo, the new McDonald's chairman, said: "I am convinced we will regain our momentum".

Figure 39 *Daily Telegraph*, 24 January 2003

in just four countries – the USA, Britain, France and Germany. There are other large multinationals, like Coca Cola, which have a much more even spread of sales around the world.

Critics of the company put its recent change of fortunes down to the following:

- Increasing competition, both in the USA and abroad. New, more health conscious fast-food restaurants are rapidly raising their market share. There are now more Subway sandwich shops in America than McDonald's outlets.
- Falling quality survey ratings against competitors.
- A poor recent record of product innovation.
- Poor locational choices for new some outlets.
- The increasing attraction of vegetarianism and scares over meat products (BSE, foot and mouth, etc.).
- The impact of critical books such as *Fast Food Nation.*

Business analysts say that unless McDonald's can respond quickly to its perceived problems, its era of rapid expansion may well be over. However, the company is investing heavily to respond to recent challenges with measures that include:

- Improving the quality of its present product range.
- Promoting higher-margin food such as the grilled-chicken flatbread sandwich.
- Getting in on the 'fast-casual' dining boom in the USA by buying chains such as Chipotle Mexican and Boston Market.

There is no guarantee that even the most powerful TNCs will maintain their market share. In fact, Starbuck's, not McDonald's, is now America's most expanionist food chain. In 2002 analysts became particularly nervous about litigation from obese Americans who blame McDonald's for their health problems. Some compared these lawsuits to the litigation that has cost tobacco companies billions of dollars. However, in January 2003 America's fast-food industry breathed a sigh of relief when a judge kicked out a lawsuit brought by a group of obese teenagers from the Bronx district of New York.

In the past McDonald's has been the target for environmental groups concerned about a variety of activities involving the company. Issues have included the destruction of rain forest, to provide grazing land for 'McDonald's cattle' and the high amount of waste packaging that has to be disposed of.

McDonald's is not only an important global entity in its own right. It has also given its name to a process or set of principles (McDonaldisation) that is said to have spread around the world. According to Ritzer these principles are:

- Efficency: McDonaldisation compresses the time span and the effort expended between a want and its satisfaction.

- Calculability: it encourages calculations of costs of money, time and effort as the key principles of value on the part of the consumer, displacing estimations of quality.
- Predictability: it standardises products so that consumers are encouraged not to seek alternatives.
- Control of human beings by the use of material technology: this involves not only maximal deskilling of workers but control of consumers by means of queue control barriers, fixed menu displays, limited options, uncomfortable seats, inaccessible toilets and 'drive-through' processing.

CASE STUDY: WAL-MART: THE LARGEST GLOBAL CORPORATION

Wal-Mart, the world's largest retailer, is expanding rapidly. With worldwide sales of $245 billion in 2002 (profits of $8 billion) it could well double this figure in five years. With over 3200 stores in the USA it is aiming in particular to be a strong presence in the world's top 20 countries, which account for 60% of all retail activity. It has been estimated that between 1995 and 1999, 25% of the US economy's productivity gains came from efficiencies at Wal-Mart. It is expected that 800,000 new jobs will be created by Wal-Mart in the USA between 2003 and 2008. Guided by founder Sam Walton's passion for customer satisfaction and 'Every Day Low Prices', the company is organised into four retail divisions: Wal-Mart Supercentres, Discount Stores, Neighbourhood Markets and SAMS CLUB warehouses. Some of the most significant stages in the expansion of the company are:

- The first Wal-Mart store opened in 1962 in Rogers, Arkansas.
- In 1968 the company expanded outside its 'home state' by starting outlets in Missouri and Oklahoma.
- In 1972 Wal-Mart was listed on the New York Stock Exchange.
- In 1977 Wal-Mart makes its first acquisition, 16 Mohr-Value stores in Michigan and Illinois.
- In 1979 it became the first company to reach $1 billion in sales in such a short period of time. By now there were 276 stores in 11 states employing 21,000 people or 'associates' as they are called by Wal-Mart.
- Expansion continued at a rapid pace and by 1985 the company boasted 882 stores with sales of $8.4 billion and 104,000 employees.
- In 1987 the Wal-Mart Satellite Network (the largest private satellite communication system in the USA) was completed, linking all operating units and the General Office.

- First supercentre opened in Washington, Missouri in 1988.
- In 1990 Wal-Mart became America's largest retailer.
- In 1991 the company opened its first foreign outlet, in Mexico City. Expansion into Puerto Rico (1992), Canada (1994), Hong Kong (1994), Brazil (1995), Argentina (1995), China (1996), Germany (1998), Korea (1998), the UK (1999) and Japan (2002) soon followed.
- Within the USA the company entered its 50th state, Vermont in 1995.
- In 1997 it became the largest employer in the USA with 680,000 employees, with an additional 115,000 workers in other countries.
- In 1997 Wal-Mart exceeded the $100 billion annual sales mark for the first time.
- By 1999 the worldwide workforce had reached 1,140,000, making the company the largest private employer in the world.
- In 2002, Wal-Mart became No.1 on the *Fortune* 500 list of the world's largest corporations, surpassing Exxon Mobil. By now the company was employing more than 1.3 million people worldwide through more than 3200 facilities in the USA and more than 1100 units abroad. More than 100 million customers per week were visiting Wal-Mart stores worldwide.

It was not until 1998 that Wal-Mart ventured into Europe when it bought Germany's Wertkauf. It reached the UK in mid-1999 with the $10.7 billion acquisition of ASDA's 229 stores. The Wal-Mart name appeared on a UK store for the first time in 2000 when the first ASDA-Wal-Mart Supercenter opened in Bristol. Now there are seven of these facilities throughout the UK. More recently Wal-Mart established a presence in Japan, its tenth operating country, buying 34% of Seiyu, a leading retailer. In China the plan is to increase the number of stores from 25 to 40 in 2003. A recent article in *Time* magazine quoted Lee Scott, Wal-Mart's Chief Executive Officer as saying 'Simply put, our long-term strategy is to be where we're not'.

As the overall size of the company has increased, Wal-Mart has reached farther back into the supply chain to source products previously bought from intermediaries. The company has opened 21 offices around the world to oversee its supplier factories. The objective is to source goods universally for all stores where feasible, so that the 250 locations in Britain and the 20 locations in Brazil can get the same price as US outlets. It is also aiming to reduce inventory expenses by speeding up the supply lines. Wal-Mart buys big – in 2002 it purchased about $6 billion worth of goods from China alone.

Its enormous size gives it a huge pricing leverage over suppliers. Wal-Mart is constantly trying to stay ahead of the competition.

Many of its standard stores are being replaced by huge supercentres. It has expanded its product range significantly in recent years and is now on the verge of moving into selling cars and banking. The company is also maximising its use of advanced technology. The most recent innovation is advanced data mining used to forecast, replenish and merchandise on a micro scale, so that even stores close to one another could offer substantially different products to match the demands of customers in their respective catchment areas. To achieve this the company has analysed every purchase made over the past ten years, looking at the relationships between the items people buy and many other variables such as time of day and price.

A key element of the Wal-Mart strategy, called 'the store of the community' is to tailor each store to the area in which it is located. The principle here is that customers differ considerably according to where they live, what they earn and many other factors. For example, in Goshen, Indiana, Wal-Mart caters to customers from the local Amish community by providing a barn where people can park their traditional horse-drawn carriages. In Shenzhen, China, store workers shout out the prices of special offers to customers accustomed to this practice in local street markets. The Wal-Mart culture also includes:

- greeters at the door
- the '10 ft rule' that insists associates must say hello to any customer that comes within that distance
- weekly staff meetings that include chants of the company name and songs
- big signs declaring 'We Sell for Less'
- minimising hotel and other costs when company executives are away on business.

Efficient air transport is an important part of company strategy, with a fleet of 20 jets kept near its Bentonville, Arkansas, headquarters. The objective is to allow managers to visit multiple markets in a single day and maintain tight central controls.

Such is the size of the company that its sales figures are as much a bellwether of America's economic health as the latest government statistics. Although the company has a very positive image at home and abroad it has not been immune from criticism:

- Wal-Mart has attracted more lawsuits that any other firm in the USA. This is largely as a result of its employment practices as it has battled aggressively to remain non-unionised.
- Civic activists accuse the company of turning CBDs into ghost towns by constructing stadium-sized superstores on the fringes of moderately sized urban areas.

Summary

- TNCs are the driving force behind economic globalisation.
- A significant proportion of world trade is intra-firm, taking place within TNCs.
- The intense global competition for market share has led to the growing domination of the largest companies in many sectors.
- The spread of a global consumer culture has been essential to the success of many TNCs.
- TNCs vary widely in size, internal characteristics and international scope.
- Large corporations often exhibit three organisational levels: headquarters, research and development, and branch plants.
- There are widely diverging views on the relative power of TNCs and nation states.
- In recent years a great variety of alliances of capital have been negotiated between TNCs.
- Investment in research and development is vital if TNCs are to maintain their market share in an increasingly fast-moving and competitive trading environment.
- There is concern that TNCs have infiltrated and manipulated important UN and other international organisations

Questions

1. **(a)** What is a transnational corporation?
   (b) Suggest why Nike (Figure 30) does not make any clothes or shoes itself.
   (c) Describe and explain Nike's cost/price chain.
2. **(a)** Use a graphical technique to compare the revenues of the world's top 12 corporations.
   (b) Give reasons why the ranking of individual corporations can change from year to year, sometimes quite dramatically.
   (c) Why are the largest companies increasingly dominating the markets in many business sectors?
   (d) Comment on the importance of branding and the mass media to the success of many TNCs.
3. **(a)** Describe the ways in which TNCs vary.
   (b) Explain the organisational levels frequently found in large TNCs.
   (c) Discuss the reasons why some national companies make the decision to go transnational.
4. **(a)** Why is there disagreement about the way in which the economic power of TNCs and countries is compared?
   (b) Discuss some of the controls countries have over the activities of TNCs.
   (c) Examine the advantages and disadvantages of TNCs to host developing countries.

6 The Role of NICs in the Changing Global Economy

KEY WORDS

Newly Industrialised Countries: nations that have undergone rapid and successful industrialisation since the 1960s.
Back-office functions: offices of a company handling high-volume communications by telephone, electronic transaction or letter. Such low- to medium-level functions are relatively footloose and have been increasingly decentralised to locations where space, labour and other costs are relatively low.

1 Different Generations of NIC

The emergence of Newly Industrialised Countries has been a key element in the process of globalisation. The term Newly Industrialised Country (NIC) is generally applied to nations that have undergone rapid and successful industrialisation since the 1960s. In Asia three generations of NIC can be recognised in terms of the timing of industrial development and their current economic characteristics. Within this region, only Japan is at a higher economic level than the NICs (Figure 40) but there are a number of countries at much lower levels of economic development. The latter form the least developed countries in the region and most, if not all, also merit this label in global terms. Although economic advance is beginning to take place in some, such as Vietnam, it is still at a very low level.

Nowhere else in the world is the filter-down concept of industrial location better illustrated. When Japanese companies first decided to locate abroad in the quest for cheap labour, they looked to the most developed of their neighbouring counties, particularly South Korea and Taiwan. Most other countries in the region lacked the physical infrastructure and skill levels required by Japanese companies. Companies from elsewhere in the developed world, especially the USA, also recognised the advantages of locating branch plants in such countries. As the economies of the first generation NICs developed, the level of wages increased resulting in:

- Japanese and Western TNCs seeking locations in second generation NICs where improvements in physical and human infrastructures now satisfied their demands but where wages were still low.
- Indigenous companies from the first generation NICs also moving routine tasks to their cheaper labour neighbours such as Malaysia and Thailand.

With time, the process also included the third generation NICs, a significant factor in the recent very high growth rates in China and

Level	Countries	GNP Per Capita* 2000
1.	Japan (an MEDC)	$ 37,676
2.	First Generation NICs e.g. South Korea	$ 9,660
3.	Second Generation NICs e.g. Malaysia	$ 3,840
4.	Third Generation NICs e.g. India	$ 485
5.	Least Developed Nations e.g. Vietnam	$ 400

*At market exchange rates.

Figure 40 Asia: five levels of economic development

India. The least developed countries in the region, nearly all hindered by conflict of one sort or another at some time in recent decades, are now beginning to be drawn into the system. It should not be too long before the economic journals recognise a fourth generation of NICs in Asia. Vietnam would seem to be a prime candidate for such recognition.

2 First Generation NICs

What were the reasons for the phenomenal rates of economic growth recorded in South Korea, Taiwan, Kong Hong and Singapore from the 1960s? What was it that set this group of 'Asian Tigers' apart from so many others? From the vast literature that has appeared on the subject the following factors are usually given prominence:

- A good initial level of hard and soft infrastructure providing the preconditions for structural economic change.
- As in Japan previously, the land-poor NICs stressed people as their greatest resource, particularly through the expansion of primary and secondary education but also through specialised programmes to develop scientific, engineering and technical skills.
- Cultural traditions that revere education and achievement.
- The Asian NICs became globally integrated at a 'moment of opportunity' in the structure of the world system, distinguished by the geostrategic and economic interests of core capitalist countries (especially the USA and Japan) in extending their influence in East and South-east Asia.

- All four countries had distinct advantages in terms of geographical location. Singapore is strategically situated to funnel trade flows between the Indian and Pacific oceans, and its central location in the region has facilitated its development as a major financial, commercial and administrative-managerial centre. Hong Kong has benefited from its position astride the trade routes between North-east and South-east Asia, as well as acting as the main link to the outside world for south-east China. South Korea and Taiwan were ideally located to expand trade and other ties with Japan.
- The ready availability of bank loans, often extended at government behest and at attractive interest rates, allowed South Korea's chaebol in particular to pursue market share and to expand into new fields.

As their industrialisation processes have matured, the NICs have occupied a more intermediate position in the regional division of labour between Japan and other less developed Asian countries.

3 Recent Economic Problems

After decades of impressive growth a major economic crisis hit the economies of East and South Asia in 1997–8, which has had a ripple effect around the world. The detail of this crisis is considered below in the case study of South Korea. The main concerns of the developed world during this crisis, which did not develop quite as badly in a global sense as some commentators feared, were:

- Japan, South Korea and others might cut the price of exports to 'jump-start' their economies and set off a trade war.
- Banking collapses could pull back large amounts of capital from the USA and the EU.
- The potential fall in demand for the developed world's exports

The economist Patrick Minford observes that the emerging economies have flooded the world market with manufactured products in wave after wave for several decades, which has produced a considerable decline in the relative price of their goods on the world market. Minford muses 'Could it be that we are observing the latest lurch in the price chart, as these countries unload their rapidly rising productive capacity on to the world market? If so, it means that, cheap as these countries are, the goods in which their capacity has been built up are in over-supply'. If this assertion is correct it will lead to falling relative wages for unskilled workers everywhere and further deindustrialisation, with manufacturing contracting everywhere in the West.

CASE STUDY: SOUTH KOREA – FIRST GENERATION NIC

Few countries have grown so rich in such a short time as South Korea. After the Korean War of 1950–3 the South had a GDP per head on a par with much of Sub-Saharan Africa and prospects that seemed no better. But from the early 1960s the economy took off, achieving startling rates of growth for the best part of four decades. Wages rose steadily and virtually all aspects of the quality of life improved. For example, life expectancy increased from 47 years in 1955 to 75 years in 2002.

Much has been written about the reasons for South Korea's success. An article in the *Economist* (29 November 1997) summed up the most important factors; 'A mixture of hard work, rigorous schooling, state-enforced austerity and imported technology transformed the economy. State-directed bank loans at negative real rates of interest allowed "strategic" industries to invest and expand at a sizzling pace'.

A prime objective was to be keenly competitive on the world market and to achieve and maintain a high volume of exports, which grew from $33 million in 1960 to $206 billion in 2000. Growth was so impressive that in December 1996 the country joined the Organisation for Economic Co-operation and Development (OECD), the club of the world's richest nations.

Huge business conglomerates, known as *chaebol*, came to dominate the economy. In 1997 the top four (Hyundai, Daewoo, LG and Samsung) accounted for over half of the country's exports. The exporting success of the chaebol encouraged them to diversify. Those named above were in an average of 140 different businesses a piece before the economic crisis. When growth rates were high such diversification was seen as a sign of strength. The military-run governments of the 1960s and 1970s held the growth of wages well below that of productivity by banning most trade union activities. In return workers were afforded excellent job security by law. A system of subcontractors, similar to Japan's, developed as an integral part of the economic system. The maxim was to invest heavily and copy the developed world's technology.

The 1997–8 Crisis

The economic crisis of 1997 and 1998 affected all the significant Pacific Rim economies to varying degrees. The South Korean economy was particularly badly hit because of the following reasons according to the financial journals:

- In late 1995 and early 1996 there was a considerable downturn in the semiconductor, metals and petrochemical businesses. At the same time the value of the yen fell, increasing the relative price of South Korea's products compared with Japan. Profits slumped and company borrowing rose. South Korea's foreign debt doubled between 1995 and 1997.

- The emergence of low-cost competitors in the region, particularly China, undercut a range of South Korean products in overseas markets.
- Manufacturing wages in 1997 were 30% higher than in Britain and because of the country's employment protection laws the industrial workforce, according to one estimate, was almost 10% larger than necessary.
- The extremely high rate of borrowing by the chaebol could be serviced when profits were high but became an enormous burden as the country's manufacturing competitiveness was eroded. As growth slowed, South Korea, under international trading pressure was forced to open its economy to a greater extent to foreign competition, which further eroded profits.
- Profligate lending by the banks tempted the chaebol to diversify into areas where they had little expertise.
- The banking system collapsed under a mountain of bad debt. One estimate in late 1997 was that: (a) 18% of the banks' outstanding loans could never be repaid; (b) 25 of the top 30 chaebols had debt-to-equity ratios of more than three-to-one.
- The heavy control of the economy by government stifled the emergence of high-quality business talent in key sectors of the economy
- The chaebols have crowded out small firms leaving the country with few innovative start-ups.

Many South Korean companies suffered considerably because of the financial crisis. In November 2000, the Daewoo car manufacturer declared bankruptcy, after billions of dollars of public funds had been spent trying to rescue the company. After talks with a number of foreign car-makers Daewoo was taken over by General Motors.

The Future

Because so many variables are involved it is difficult to estimate how long the restructuring of the economy will take and what the total cost will be. Whether South Korea likes it or not the chaebol will be exposed to more competition from abroad in the future. Although painful in the short term, they will be forced to improve efficiency and productivity to survive if they are to fend off the challenge from countries with cheaper labour such as Thailand and China, and from those with better technology and greater economies of scale such as the USA and Japan. However, South Korea's strong reliance on heavy industry could make its problems harder to resolve compared to economies based more firmly on light manufacturing and services.

More and more South Korean companies are moving production to China to take advantage of lower costs. For example, the LG group of companies invested $1.5 billion in production in China between 1992 and 2002. Samsung has also built plants in China, with Hyundai soon to follow. However, it is not just lower labour and other costs that are attracting South Korean companies to China. The latter is becoming a more and more important market for Korean goods.

Then there is the unpredictability of neighbouring North Korea. There is always the risk of war but many observers think that North Korea will collapse, as East Germany did, and then demand to be reconstructed with South Korean help. If this were to be the case the financial burden would be immense.

CASE STUDY: INDIA – A THIRD GENERATION NIC

The primary sector still dominates employment in India. About three-quarters of the population are engaged in this sector of the economy, which accounts for around 25% of GDP. However, the majority of landholdings are farmed at subsistence level resulting in a very high degree of rural poverty. In 1999 44.2% of the population of India lived on less than $1 a day.

Although India has a reasonably diverse manufacturing base it has not as yet achieved the rapid expansion experienced by Asian rivals such as South Korea, Taiwan, Thailand, Malaysia and China. This has been at least partly due to a low level of foreign direct investment, a situation that has begun to change in recent years with the introduction of a number of important economic reforms.

The transformation of the Indian economy has, in fact, been led by the service sector. The country has a large number of highly qualified professionals whose skills are in demand in other countries, particularly in the English-speaking world. Demand and supply have been united by telecommunications and by worker migration.

A milestone in India's economic progress was reached in 1999 when Infosys Technologies, a software exporter, became the first Indian company to list on a US stock market. It was joined by ICICI, the first Indian financial institution to undergo a US audit, and Satyam Infoway, an internet service provider. In the same year Tata Tea tabled a record bid for a foreign company, the UK's Tetley Tea, while software exports hit a new record. The IT sector's market capitalisation exceeded Rs 1000 billion: the highest valuation ever placed on an Indian industry.

(a) India: A Different Route to NIC Status

India has achieved economic growth via a different route to its prominent Asian economic rivals. This is because:

- The rapid economic growth of the 1990s was due much more to the expansion of the service sector than to manufacturing (Figure 41). Services accounted for over 47% of GDP in 2000. Foreign input into manufacturing has, until recently, been only in the form of joint ventures due to legal constraints.
- The filter-down of employment to India has been dominated by business links with North America and Europe, with a relatively low level of involvement from Japan and the other dynamic Asian economies.

The service sector is growing rapidly across the board and not just in software and ICT services. Media, advertising, retail, personal financial services, entertainment, tourism and leisure have all expanded at a significant pace in the last decade. For example, Indian media companies export content to ethnic markets around the world.

(b) The Slow Growth of India's Manufacturing Sector

In the decades after Independence (1947) India followed a policy of *import substitution* (manufacturing its own products rather than importing from other countries), which led to the development of a broad industrial base. Under this policy (which was designed to reduce by as much as possible the amount India spent on imports), Indian industry was heavily protected from foreign competition. Thus, until the economic reforms of the early 1990s this drive for self-sufficiency took place with a minimum of foreign participation. However, the lack of competition

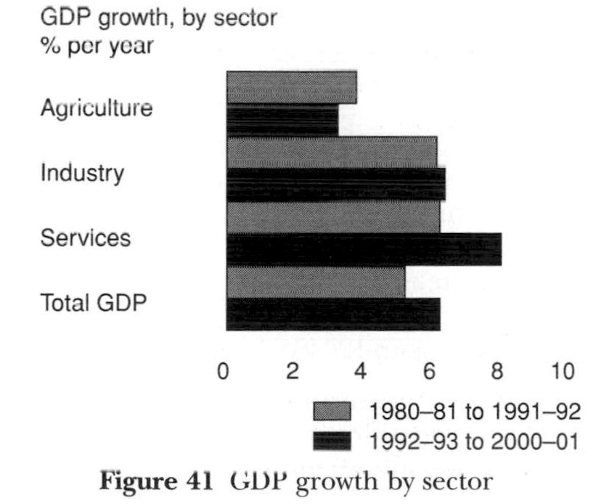

Figure 41 GDP growth by sector

contributed to poor product quality and inefficiencies in production. Indian manufactured products were often viewed as poor quality in the domestic market and for the same reason made little impact abroad. Most people would struggle to name an Indian manufactured product that is widely sold in Britain or in other developed countries.

In the early stages of economic reform the only way foreign companies were allowed to enter the manufacturing sector in India was through joint ventures (shared ownership between foreign and Indian companies). Thus, many of India's 15,000 joint ventures were formed for regulatory as opposed to strategic reasons. As a result of such regulatory control India did not benefit from the level of manufacturing filter-down from developed countries that other Asian NICs experienced. However, since 1999 100% foreign ownership has been allowed in many industrial sectors. Hero Honda, the motorcycle manufacturer, is India's most successful joint venture. Honda, which has a 26% stake in the company, announced in 1999 that it had secured permission to set up a 100% owned subsidiary after signing a five-year non-competition agreement. The move to 100% foreign ownership is a significant reform for India. The government hopes that it will lead to a surge of foreign direct investment, following the pattern set by the most successful Asian economies. Yet, it is not surprising that many Indian companies, used to a relatively high level of protection from foreign competition, are uneasy.

Signs of progress in the manufacturing sector include the following:

- Madras (Chennai) has become a focal point for the car industry with both Hyundai and Ford established there. Since 1991 the number of car manufacturers in India has increased from three (190,000 vehicles) to ten (500,000 vehicles) in 2000.
- The western states of Gujarat and Maharastra are building on their heavy industrial base and natural resources to built modern ports. These two states combined attract about a quarter of all new investment, both domestic and foreign, into the Indian economy.
- Reliance, the oil and petrochemicals group, commissioned the world's largest integrated oil refinery in 1999, at 27 million tonnes per annum. Reliance is at the forefront of advancing vertical integration in India, spanning the value chain from crude oil to textiles.
- The average energy consumption in the cement industry per tonne of output fell from 120 kW hours in 1995 to 85 kW in 1999.

As a result of changes in government policy, foreign direct investment rose from an insignificant level in 1990 to over $2 billion a year in 1999. The government aims to increase FDI to $10 billion a year by 2005. By way of comparison, China is currently attracting $40 billion a year in FDI.

(c) Economic Reforms

The more open India's economy is the more likely that foreign investors will be enticed to invest in it. Although India has successfully undergone a 'first wave' of economic reforms it needs to implement a 'second wave' of reforms to bring it into line with the more successful economies of the region. The first tentative liberalisation of India's economy began in 1991. These reforms lifted national output to an annual growth rate of nearly 7% before slipping back to just over 5% in the late 1990s. According to a recent *Financial Times* survey 'Until India radically transforms its economy it will not reach the pre-crisis (1997–8) East Asian growth levels of 8–10% of GDP a year, which alone will raise its 400 million people from the sub-subsistence of living on less than $1 a day'.

The most immediate problems to tackle are the large budget deficit (the combined deficit of central government and the states is about 8.5% of GDP) and financial sector reform to unlock the big and growing pool of private savings. A budget deficit occurs when the government spends more than it receives from taxation and other sources of revenue. The problem with a large budget deficit is that the government has to borrow to finance it. The two main implications of this are:

- After interest payments and other costs associated with running a large deficit the government is left with little money to invest in the economy directly. Health, education and infrastructure are all areas where new investment is required to raise pitifully low standards.
- A high level of government borrowing leaves little money left for the private sector to borrow. This is known as the 'crowding out' of the private sector.

At present India is in a strong position with regard to savings, with a long run of good monsoons and rising rural incomes. The National Council for Applied Economic Research (NCAER), a reputable Delhi think-tank, sees the saving ratio (the proportion of income that is saved) reaching 36% of GDP – surpassing East Asian levels – by 2006 if it continues to rise at the rate of the past decade. Modern economists, following WW Rostow in the 1960s, see this as a very powerful engine of growth.

While the service economy powers forward many traditional manufacturing industries are showing signs of strain. A major challenge for India as it introduces new reforms to its economic system is to enable traditional industry to restructure fast enough to still have a role in the emerging new economy. The country's poor infrastructure is a major handicap for many manufacturing sectors.

(d) The Software and ICT Services Sector

India's software and ICT services sector has been at the forefront of the country's economic growth over the last decade. Revenues from Indian software exports grew from less than $1 billion in 1996 to $4 billion in 2000. Currently accounting for 2% of GDP, it is estimated that this sector will contribute 7% by 2008. Indian ICT expertise operates both at home and abroad. It is well represented in America's Silicon Valley, the City of London and many other high-technology centres in the developed world.

India's ICT sector has benefited from the filter-down of business from the developed world. Many European and North American companies, which previously outsourced their ICT requirements to local companies, are now using Indian companies. Outsourcing to India occurs because:

- labour costs are considerably lower
- a number of developed countries have significant ICT skills shortages
- India has a large and able English-speaking workforce (there are about 50 million English-speakers in India).

The main factors limiting the degree of filter-down of ICT to India are the management systems and overheads involved. Because of these important factors it is likely that the majority of 'offshore' outsourcing will focus on maintaining legacy code (well-established computer programs) while new developments using Indian expertise will involve skilled labour migrating to the centres of demand in the developed world. To combat ICT skills shortages countries such as the UK have speeded up the process of issuing work visas. In the UK the process is now down to between five and ten working days.

In 1999 India established a Ministry of IT. The Ministry's main task is to increase software and ICT services revenue for the country. From low-technology beginnings, Indian companies are migrating to high-value software services, e-commerce, business consultancy and technology research. This is all to maximise India's real cost advantage, which is in brainpower and not manpower.

Bangalore, Hyderabad and Madras, in the south, along with the western city of Pune and the capital city Delhi, have emerged as the centres of India's ICT industry. The country's biggest ICT companies, such as Infosys, Wipro and Satyam, have all developed from these dynamic clusters of industry. The early growth of the sector was driven by foreign companies, mainly American. It has only been relatively recently that the number of home-grown companies has significantly increased. Texas Instruments, which has had facilities in Bangalore since 1985 saw half a dozen of its top Indian engineers leave the company in 2000 to strike out on their own.

Over the past decade India has exported its ICT expertise to many other counties. However, it is possible that India itself may suffer ICT manpower shortages in the future. According to a recent survey by India's National Association of Software and Services companies, demand for IT professionals is expected to rise from 340,000 in 2000 to 800,000 by 2005. Major companies are very much aware of this possibility. For example, Cisco has pledged the Indian government $10 million to set up 30 'networking academies' across the country.

The Indian government hopes that Indian firms and individuals who have made big money abroad, or by doing contract work for foreign companies in India, will invest in fledgling Indian enterprises in the form of venture capital. This process has already begun and could become very significant indeed in the coming years.

The recent integration of India into the world economy has witnessed the emergence of a large number of companies that did not exist ten years ago. In 1995 the IT sector was not represented in India's top ten private sector companies. By 1999 four IT companies had joined this list.

(e) Back-office Functions

A number of the financial journals describe India as the 'back office of the world'. Such functions go well beyond ICT services to encompass a wide range of office skills that richer economies are only to eager to outsource to much lower wage nations. The term 'IT-enabled services' is often used to describe such functions. NASSCOM, India's main association of information technology companies, estimates that India will employ 1.1 million people and earn $17 billion from IT-enabled services by 2008. A report to the Electronics and Computer Software Export Promotion Council, a government body, sees the industry's exports to America growing from $264 million in 2000 to over $4 billion in 2005.

India's back-office industry has two sections:

- 'Captive' operations of large Western companies seeking to cut back-office costs without outsourcing. For example, GE Capital Services opened India's first international call centre in the mid-1990s, which now employs over 5000 people. Other companies that operate major back-office facilities in India include American Express, British Airways and Swissair. Western companies often save 40–50% by shifting work to India. Call

centres often give their staff American or European pseudonyms, a practice that has become something of a national joke.
- 'Shorter-term' contracts between western companies and subcontractors in India, often brokered by middlemen. An important element in this sector is medical transcription, in which companies convert dictation by doctors in America into written records. India has about 200 medical-transcription companies employing 10,000 transcribers. Intense competition between them has driven down costs, which may attract other developed countries to make greater use of such services.

India's teleworking industry can be divided into five sectors:

- Data entry and conversion, for example medical transcription.
- Rule-set processing. Here a worker might decide, under an airline's rules, whether a passenger qualifies for an upgrade.
- Problem-solving, for example deciding if an insurance claim should be paid.
- Direct customer interaction. Here the teleworker handles more elaborate transactions with the client's customers.
- Expert 'knowledge services', which require specialists (e.g. engineers and lawyers) using databases.

(f) The Regional Consequences of Economic Growth

A worrying side-effect of growth in the national economy is the widening regional development gap between north and south. Most investment is flowing southwards into a V-shaped region from Gujarat in the west to Andhra Pradesh in the east, into those states at the forefront of economic reform. The Indian government hopes that the obvious success of the southern states will give the northern states the impetus and confidence to undergo economic change.

Gujarat and Maharastra recorded the fastest growth rates in the 1990s. Together these two states attract about a quarter of all new investment in India, both domestic and foreign. The southern states of Andhra Pradesh, Karnataka and Tamil Nadu attract another 22%. In contrast, in half of the 14 biggest states growth rates actually declined in the 1990s. In 1980, Uttar Pradesh, Bihar and Orissa accounted for 38% of India's poor. That share rose to 46% in 1999. There is a general perception that the coastal states have always been more outward looking than land-locked states, which tend to be much more traditional in both social and economic terms.

The models of regional development constructed by Gunnar Myrdal (1957) and Albert Hirschman (1958) explain that in the early stages of economic development the wealth gap between core and peripheral regions widens. Such a condition is referred to as regional economic divergence. The rapid growth of the

Indian economy in the 1990s has followed these models with the gap between rich and poor states expanding steadily. Only when a country has reached a stage of economic maturity does regional economic convergence usually occur.

Analyses of regional growth have concluded that:

- There is a strong link between levels of investment and the quality of certain kinds of infrastructure, particularly electricity and telecommunications.
- Higher growth states tend to have higher private (not public) investment.
- Investors were wary of states that were poor, badly governed and racked by inter-caste violence.

The south and west have better educational standards, greater longevity and smaller families than the national average. In addition, more equitable landholding patterns in rural areas make income disparities less pronounced than in poorer states such as Uttar Pradesh and Bihar.

(g) Conclusion

The growth rate of the Indian economy over the past decade has been impressive, but the benefits of development have as yet only affected a minority of the population. Although India is now a global player in ICT, the country remains one of the least networked societies in the world with fewer than five million computers. As Krishna Guha states (*Financial Times*, 19 November 1999) 'What is unclear, however, is whether the technology revolution will transform India's domestic economy or whether India's digital economy will remain a virtual outpost of the USA, staffed by the English-speaking middle class, unconnected to its geographical surroundings'.

While there are many promising signs for the economy as a whole, formidable obstacles to more far-reaching economic development remain. These include:

- Inefficiencies in the banking sector.
- A high level of government borrowing.
- Manufacturing industries that have yet to become fully consumer-oriented because of decades of government protection.
- Labour laws that make it difficult for firms to respond rapidly to changing market conditions.
- A poor level of hard and soft infrastructure.
- The relatively low efficiency of publicly owned services.
- A high level of illiteracy.

Summary

- Newly Industrialised Countries have developed since the 1960s.
- The first group of NICs, the so-called four Asian Tigers, were South Korea, Taiwan, Singapore and Hong Kong.
- The rapid economic growth of these particular countries was due to geographical location, levels of education, standard of infrastructure, the availability of capital and a 'moment of opportunity' in the structure of the world system.
- Today, in Asia, three generations of NIC can be recognised.
- Manufacturing industry has filtered down from one generation of NIC to another.
- The financial crisis of 1997–8 was the most serious interruption that has occurred to the growth of South Korea and other NICs. This crisis had significant consequences including bankruptcy, unemployment and industrial restructuring.
- India has achieved economic growth via a different route to its prominent Asian economic rivals. India's rapid economic growth has been due more to the expansion of the service sector than to manufacturing.
- India is now a global player in ICT.
- A worrying side-effect of growth in the Indian economy is the widening regional development gap.

Questions

1. **(a)** Define the term 'Newly Industrialised Country'.
   (b) Compare the GNP per capita of the three generations of NIC illustrated in Figure 40.
   (c) Discuss the reasons for the development of the first generation of NIC.
   (d) Explain the development of subsequent generations of NIC in Asia.
2. **(a)** Examine the reasons for South Korea's rapid economic advance since the 1960s.
   (b) Why was South Korea so badly affected by the recent global economic crisis?
   (c) How and why is the economy of South Korea likely to change in the future?
3. **(a)** How has recent economic growth in India differed from the economic growth of countries such as South Korea and Taiwan?
   (b) Why was economic reform so important in attracting foreign investment into India?
   (c) Examine the growth of the ICT sector in India.
   (d) Discuss the regional consequences of economic growth.

7 Least Developed Countries

KEY TERMS

Least Developed Countries: the poorest and weakest economies in the developing world. LDCs are a subset of the LEDCs
OECD: The Organisation for Economic Co-operation and Development set up in 1961 – a grouping of the more economically developed nations of the world.

> The test of our progress is not whether we add more to the abundance of those who have much; it is whether we provide enough for those who have too little.
>
> *Franklin Delano Roosevelt*

While the Newly Industrialised Countries are an important part of the successful side of globalisation the Least Developed Nations are on the other side of the coin, having been largely by-passed by the processes of wealth creation.

1 Identifying LDCs

The concept of Least Developed Countries was first identified in 1968 by the United Nations Conference on Trade and Development (UNCTAD). It was subsequently updated by the UN to describe the 'poorest and most economically weak of the developing countries, with formidable economic, institutional and human resource problems, which are often compounded by geographical handicaps and natural and man made disasters'. LDCs are a subset of the LEDCs. Twenty-four LDCs were identified in 1971 but by 2001 that number had grown to 49 (Figure 42). With 10.5% of the world's population (610.5 million) these countries generate only one-tenth of 1% of its income. The list of LDCs is reviewed every three years by the Economic and Social Council (ECOSOC).

The criteria used to identify LDCs have changed over time and become more complex. The current criteria are:

- A low income, as measured by GDP per capita.
- Weak human resources, as measured by a composite index (Augmented Physical Quality of Life Index) based on indicators of life expectancy at birth, per capita calorie intake, combined primary and secondary school enrolment, and adult literacy.
- A low level of economic diversification, as measured by a composite index (Economic Diversification Index) based on the share of manufacturing in GDP, the share of the labour force in industry,

annual per capita energy consumption, and UNCTAD's merchandise export concentration index.

UNCTAD uses different thresholds for inclusion in, and graduation from, the LDC list. A country qualifies for entry to the list if it meets inclusion thresholds on all three criteria. A nation qualifies for graduation from the list if it meets thresholds on two of the three criteria. In terms of low income the current threshold for inclusion on the LDC list is $800 per capita, while the threshold for movement off the list is $900 per capita. In its July 2000 review ECOSOC declared that Senegal was eligible for LDC status (subject to the government of Senegal so desiring) and decided to postpone until 2001 its consideration of the Maldives graduation from the LDC list.

The criteria for determining LDC status are currently under review. The Committee for Development Policy has recommended that the Economic Diversification Index be replaced by an Economic Vulnerability Index reflecting the main external shocks to which many low-income countries are subject.

2 Three UN Conferences

The first United Nations Conference on the Least Developed Countries was held in Paris in 1981. However, despite major policy reforms initiated by many LDCs and supportive measures taken by a number of donor countries in the areas of aid, debt and trade, the economic situation of these countries as a whole worsened in the 1980s. Thus, a second UN Conference on the Least Developed Countries was held in Paris in 1990. The outcome of the Conference was embodied in the Paris Declaration and the Programme of Action for the Least Developed Countries for the 1990s. The third UN Conference on the Least Developed Countries was held in Brussels in May 2001.

3 The Income Gap Widens for LDCs

Income is a very important means of enlarging people's choices and is used in the UN's Human Development Index as a proxy for a decent standard of living. Income levels across countries worldwide have been both diverging and converging. In 1960 there was a bunching of regions, with East Asia and the Pacific, South Asia, Sub-Saharan Africa and the Least Developed Countries having an average per capita income around one-ninth to one-tenth of that in high-income OECD countries. However, since then the LDC's have fared worse than any other region, deteriorating to about one-eighteenth of incomes in high-income OECD countries.

The NGO (Non-governmental Organisations) Forum at the Brussels Conference stated that 'Globalisation according to the free market model is making the rich richer and the poor poorer'. The causal factors identified were:

- The World Trade Organisation has undermined the interests of LDCs.
- Global ODA has never reached the level of the UN Commitments of 0.7% of GNP and 0.15% to LDCs.
- Initiatives to cancel debt have advanced too slowly with too little effect. Many LDCs spend 40% of their GDP on debt servicing.
- In many countries development has been held back or put into reverse by the impact of HIV/AIDS and conflicts.

4 Four Categories of LDC

The Least Developed Countries 2000 Report identifies four subgroups within the LDCs according to growth and instability of GDP per capita 1990–1998 (Figure 42). These are:

- Group 1 countries – where the real GDP per capita growth exceeded 2% per annum. This is a rate at which their incomes are converging with average developing country performance.
- Group 2 countries – where per capita income is growing but where incomes are regressing relative to average performance in the developing countries.
- Group 3 countries – where per capita income is regressing in absolute terms at less than 2% per annum.
- Group 4 countries – where per capita income is regressing in absolute terms at more than 2% per annum.

Seven of the Group 1 LDCs are in Asia. Of the 22 countries in Groups 3 and 4, 11 have experienced serious armed conflicts and internal instability during the 1990s.

The instability index is measured as the standard deviation of annual growth rates over the 1988–98 period. An important feature of the performance of most LDC economies is the significant degree of income instability. For example, Equatorial Guinea, which experienced by far the highest average growth rate between 1990 and 1998, also had the highest rate of income instability.

5 Major Problems of LDCs

As the gap between the richest and poorest countries in the world widens LDCs are being increasingly marginalised in the world economy. Their share of world trade is declining and in many LDCs national debt now equals or exceeds GDP. Such a situation puts a stranglehold on all attempts to halt socio-economic decline. Figure 43 shows how long it will take individual LDCs to reach the threshold of

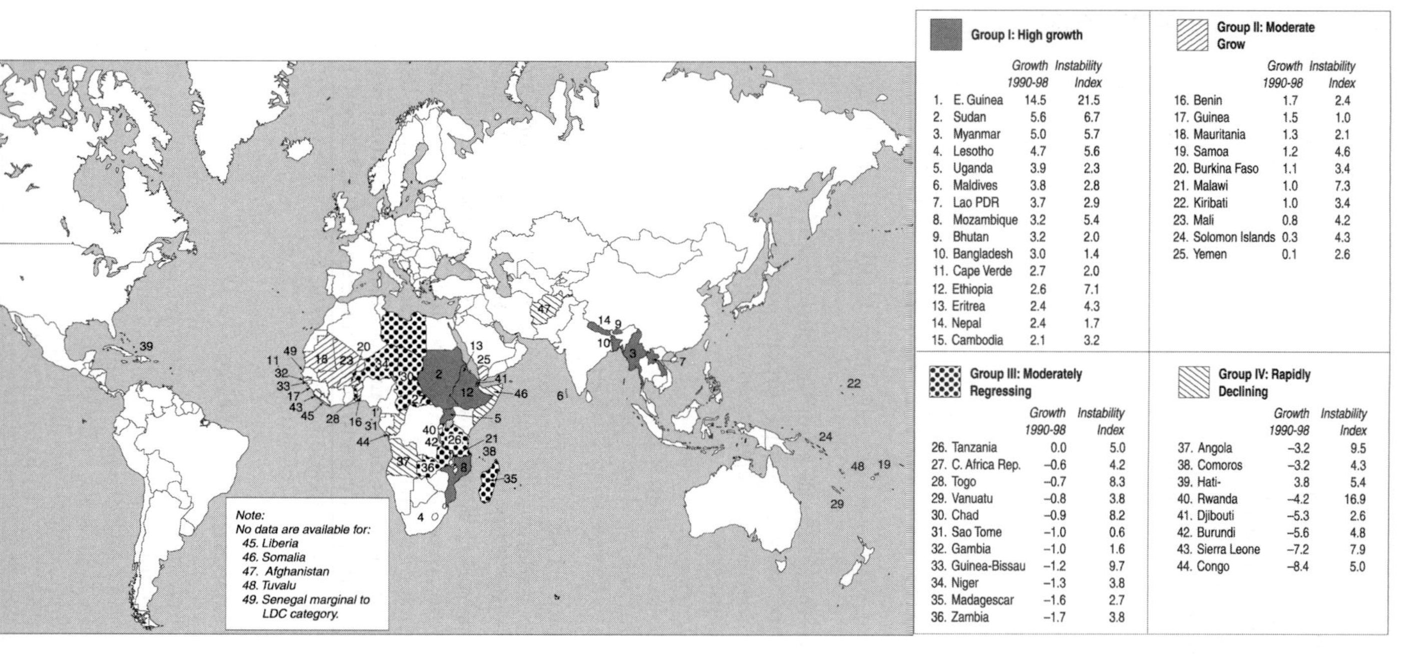

Figure 42 Statistical profile of LDCs, 2001

Already there	18–25 years	25–50 years	50–100 years	>100 years	Negative growth or stagnant
Cape Verde	Bhutan	Bangladesh	Benin	Burkina Faso	Angola
Equatorial Guinea	Lao PDR	Guinea	Cambodia	Malaŵi	Burundi
Maldives	Lesotho*	Mozambique	Eritrea	Mali	Chad
Vanuatu	Sudan	Uganda	Ethiopia	Yemen	Comoros
			Mauritania		Congo
			Nepal		Gambia
					Guinea-Bissau
					Haiti
					Madagascar
					Niger
					Rwanda
					São Tomé
					Sierra Leone
					Solomon Is.
					Togo
					Tanzania
					Zambia

Notes: The $900 income target is set at 1997 US $. The base year for calculations is 1997. Projections are based on trend growth rates of 1990–8.
**Lesotho reaches $900 threshold in 15 years, all other countries in this group of countries are above 18 years.*

Figure 43 How long will the LDCs take to reach $900 per capita income levels if curent trends persist?

$900 per capita income (for graduation off the LDC list) if current trends persist.

Although life expectancy at birth is rising, the rate of increase slowed in the 1990s with the gap between LDCs and other developing countries remaining stubbornly wide. This is perhaps not surprising when health expenditure per capita is examined. Here, as expected, it is in the African LDCs where the health crisis is at its worst.

Education is, of course, a major factor in the development process and in particular female education has been highlighted as a catalyst for growth. The gender gap in education for LDCs remains significantly higher than for other developing countries. At just over 80% the gender gap for LDCs remained static in the 1990s.

MAURITANIA: CASE STUDY OF AN LDC

The west African nation of Mauritania, at over four times the size of the UK, covers more than 1 million km² (Figure 44). It borders Western Sahara, Algeria, Mali, Senegal and the Atlantic Ocean. Water is a prime concern not just in the desert and semi-desert areas but also in those zones where rain-fed agriculture is possible. Agricultural production varies hugely from year to year because of the unpredictability of the rains.

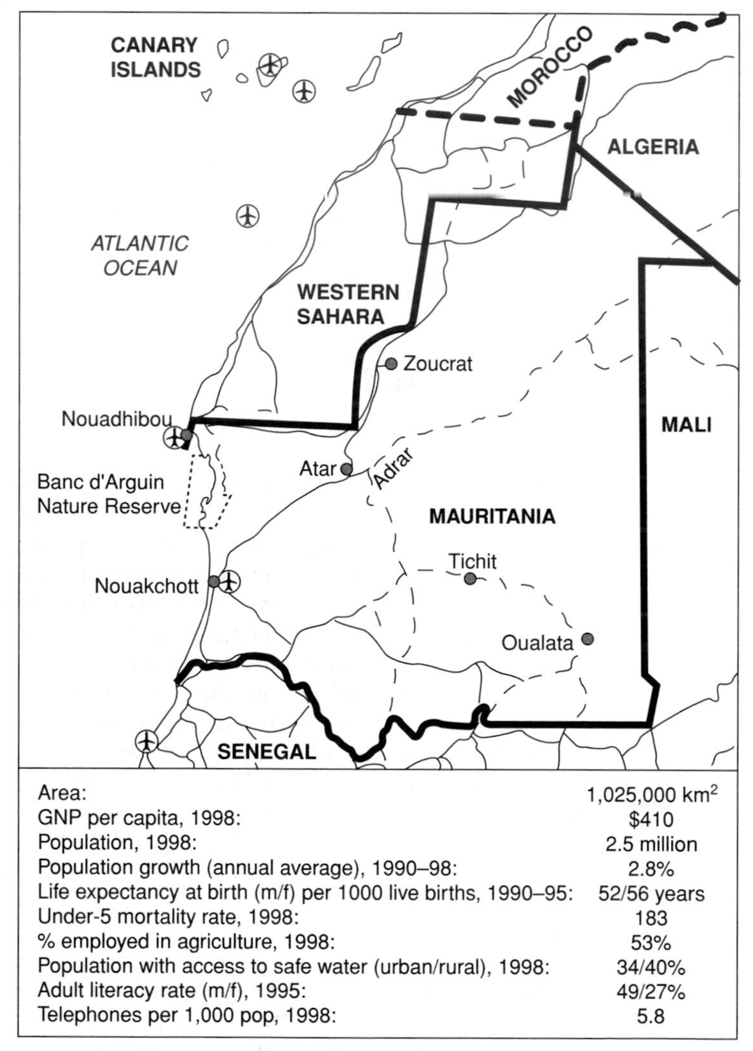

Area:	1,025,000 km²
GNP per capita, 1998:	$410
Population, 1998:	2.5 million
Population growth (annual average), 1990–98:	2.8%
Life expectancy at birth (m/f) per 1000 live births, 1990–95:	52/56 years
Under-5 mortality rate, 1998:	183
% employed in agriculture, 1998:	53%
Population with access to safe water (urban/rural), 1998:	34/40%
Adult literacy rate (m/f), 1995:	49/27%
Telephones per 1,000 pop, 1998:	5.8

Figure 44

With a relatively small population of 2.5 million, Mauritania is classed as a Group 2 LDC (Figure 43). Its annual average growth in GDP of 1.3% between 1990 and 1998 gives some cause for optimism for the future.

When the country gained independence from France in 1960, slavery still formally existed and under 5% of the population lived in urban areas. Today the urban population numbers over 60%. In an effect to shed its poverty the country has embraced privatisation and other aspects of structural adjustment favoured by the World Bank and the International Monetary Fund. Mauritania is one of the first countries to benefit from the heavily indebted poor country (HIPC) debt reduction scheme.

Although starting from a low base there are significant signs of economic and social progress. For example:

- Between 1990 and 2000, the proportion of Mauritanians living below the poverty line fell from 56% to 43%. The objective is to further reduce this figure to 39% by 2004 and 17% by 2015.
- The gross primary enrolment ratio reached almost 90% in 2000–1. Since 1995 the gap between enrolment ratios of boys and girls has almost closed.

The Economy and Trade

Exports have depended almost totally on two commodities: iron ore and fish. In 1998, iron ore accounted for 47.8% of total exports by value. The relatively low price of iron ore on the world market and the depletion of some of the best reserves have made Mauritania's dependency on iron ore very risky. Fortunately, the fishing sector has developed strongly in recent decades but the problem of over-fishing (see below) by foreign boats is now causing considerable concern. Fishing accounts for about half of export earnings but its contribution to the country's GDP is considerably lower.

Nouadhibou is the main port of export for the mining and fishing industries. Its turnover makes it one of the most important ports in West Africa. However, it has one great disadvantage – there is no road link with the capital Nouakchott. This situation, however, will change when the West Coast pan-African highway is completed. An important development in the south was the construction of a half-a-million tonne capacity deep-water port near Nouakchott with Chinese aid. There has been talk of turning Nouakchott Port into a major gateway for landlocked Mali, which would be of considerable benefit to both countries.

Discoveries of other minerals, including gold and diamonds, offer possibilities for the future but further exploration will be required to ascertain the economics of development. Mauritania, like all oil-poor LEDCs, can do nothing but pay the fluctuating market price for its essential oil imports. However, the discovery of oil off the coast in mid-2001 may solve this problem if the reserves prove economic to exploit.

Import substitution has long been an objective but earlier attempts at establishing an oil refinery and other industrial plants ended in failure. The very small size of the domestic market has proved to be a major limitation in this respect.

The periodic need to import food is a major obstacle to achieving a trade balance. Traditionally, the nomadic desert Moors were largely self-sufficient, while in the south, the predominantly black African farmers also grew their own food. However, a high rate of rural–urban migration in recent decades has increased the number of people dependent on others to produce their food. Acute periods of drought, notably in the 1970s and 1980s, have exacerbated the situation. Rice production, the staple food of many Mauritanians, has been of particular concern. Yields can vary significantly from year to year, depending on rainfall and other factors. Thus, Mauritania is not always self-sufficient in rice. In addition, imported rice is often favoured as it tends to be less expensive. Livestock rearing accounts for 15% of GDP. There are more goats than people in Mauritania and more than a million camels.

Since independence, France has remained Mauritania's main trading partner. Migrant workers based in France remit significant sums home. In total the EU accounts for just over 60% of both imports and exports. Mauritania is a signatory to the Cotonou Agreement, which links more than 70 African, Caribbean and Pacific countries to the EU, giving them various trade and aid concessions. China and Japan have also been important trade and aid partners.

Development Programmes

(a) Sustainable Rural Development

The promotion of sustainable rural development is vital if the drift to the towns is to slow. Among the government agencies and projects involved in this process is the Programme for Managing Natural Resources in Rain-fed Zones (PGRNP). The main objective of this programme is to stem the degradation of vegetation cover and to improve the environment by enabling local populations to rationalise the use of natural resources. The approach is highly decentralised with over 280 autonomous decision-making centres. NGOs, such as the Paris-based SOS Sahel, have been working with local communities to try to conserve tree cover and replant. There has been a push to replace firewood with buthane as the main domestic fuel.

(b) The Senegal River Valley

In this region where irrigation enables far more intensive farming than elsewhere a project jointly funded by the World Bank and the government is under way. The objectives of the Programme for Integrated Development of Irrigated Agriculture in Mauritania (PDIAIM), which has been organised in three phases from 2000 to 2010, are:

- higher agricultural output
- wider crop diversification
- a reduction in rural poverty
- improved food security
- a better ecological balance.

(c) The Oasis Development Project

This project, which was established in the mid-1980s, is assisted by the UN's International Fund for Agricultural Development (IFAD) and the Arab Fund for Economic and Social Development. The objective is to improve the living standards of poor people in the oasis zones. An important part of this participatory development scheme has been the establishment of micro-credit co-operatives of which there are now about 70. Members can borrow small amounts of money, interest-free, for up to ten years for household and community self-sufficiency projects (e.g. purchasing seed, digging new wells, fencing land). When the project ends in 2002 it is hoped that private investment will come to play a more important role as the tourist value of oasis settlements is developed.

(d) Nouakchott

Nouakchott, which means 'the place of the winds', is now the fastest growing city in Africa. Previously a small French military post, Nouakchott was chosen as Mauritania's capital when the country became independent in 1960. The advantage of selecting what was largely a virgin site was that the new city could be planned from scratch. Owing to the nomadic nature of much of the nation's population at the time it was assumed that a settlement of around 50,000 people would be adequate. Today, the population is well over ten times the original estimate. As well as the usual rural–urban migration that characterises developing countries, urban growth has been exacerbated by large numbers who are fleeing drought and the advancing desert, as well as refugees from regional conflicts. The result is huge informal neighbourhoods of tents, shacks and breeze-block houses.

A ten-year urban development project funded by the World Bank is attempting to alleviate the worst deprivation in the capital. Its objectives are to improve water and electricity supply, upgrade educational facilities, and make micro-credit facilities available for the creation of small- and medium-sized businesses, particularly in the construction and service sectors.

(e) Tourism

Until the late 1990s tourism was not a priority in Mauritania because:

- There was concern that an influx of tourists would damage the nation's cultural and religious heritage.
- The huge strain on government finances could not justify investing in the infrastructure required for a significant tourist industry.

However, in recent years the government has looked more favourably on tourism in order to: (a) generate revenue; and (b) as a contributing factor in ecological and cultural preservation. The aim is to avoid mass tourism and all its pitfalls and to target special interest groups and adventure tourists. The main attractions on offer are:

- The Adrar region in the interior. Here the four towns of Chinguetti, Ouadane, Oualata and Tichitt, which date back to the 12th and 13th centuries, have been declared World Heritage sites by UNESCO. Exploration of these old Sahara and Sahel trade route cities can be combined with desert travel. Charter flights fly regularly from Marseilles to Atar.
- The countryside along the Senegal river in the south. Here the pioneering Mauritanian tourism company El Mejabat El Koubra Tours (MKT) has set up a permanent camp at Keur Macene, which has its own 120,000-hectare hunting reserve. The camp that began with 30 rooms will expand to 100 in 2003.
- The Atlantic coast near Iwik offers the sight of hundreds of thousands of migrating birds, feeding and resting in the nature reserve of the Banc d'Arguin. In this region the Imraguen people use the traditional method of getting dolphins to chase fish into their nets.
- The train running from Nouadhibou to the mining areas at Zouerat and beyond is becoming a favourite with railway buffs. The train, said to be the world's longest and which mainly consists of wagons carrying iron ore, gives a fascinating perspective into a unique landscape.

Summary

- The number of LDCs increased from 24 in 1971 to 49 in 2001.
- The criteria used to identify LDCs are low income, weak human resources and a low level of economic diversification.
- Four categories of LDC have been identified by the UN.
- LDCs are being increasingly marginalised in the world economy.
- In many LDCs development has been held back or put into reverse by the impact of HIV/AIDS and recent wars/internal conflicts.
- Mauritania is one of the first countries to benefit from the heavily indebted poor country (HIPC) debt reduction scheme.
- Mauritania and the other poor nations of West Africa are suffering from the activities of high-technology fishing fleets from developed countries, including the EU.
- Mauritania is heavily dependent on the export of primary products for foreign currency.
- The country has embraced privatisation and other aspects of structural adjustment favoured by the IMF.
- It is hoped that the various development programmes underway will raise living standards.

Questions

1. (a) Describe the geographical distribution of LDCs.
   (b) Discuss the criteria used to identify LDCs.
   (c) Explain the basis of the classification of LDCs into four groups.
   (d) Examine the major problems of LDCs.
2. (a) Describe the geographical location of Mauritania.
   (b) Outline the economic and trading situation of the country.
   (c) Discuss some of the development programmes that are underway in Mauritania.

8 The Global Economy of the Future

KEY TERMS

Religious fundamentalism: movements favouring strict observance of religious teaching (Islam, Christianity, Hinduism, etc.)
Sustainable development: development that meets the needs of the present without compromising the ability of future generations to meet their needs.
Telemobility: the ability of employees, through communications technology, to work at locations other than a centralised office.

1 Advanced Globalisation

There can be little doubt that the process of globalisation has some way to go. When it will be complete, if ever, is a source of much debate, as are the likely outcomes. Some see the process as offering enormous opportunities, while others are fearful of its extension. The literature on the subject frequently cites the following as the consequences of advanced globalisation:

- The elimination of geography as a controlling variable in the global economy.
- The disappearance of the nation state.
- Economic synchronisation across the globe.
- Companies with no specific territorial location or national identity.
- The disappearance of distinctions between LEDCs and MEDCs as structures of wealth and poverty became detached from territory.
- English as the common public language of the globalised system.

According to Waters: 'The rise in global consciousness, along with higher levels of material interdependence, increases the probability that the world will be reproduced as a single system'. Although there has been a strong movement towards a single system in recent decades, the degree of conflict around the world emphasises that there is a lack of agreement on what shape the single system should take in the future. Robertson argues that globalisation is neither necessarily a good nor a bad thing but that its moral character will be determined by the people of the planet. Since the Seattle demonstration in 1999 the level of concerned debate about the best way forward has increased. Fewer people are now prepared to leave it just to governments and international economic organisations to decide. Professor Benjamin Barber, director of the Walt Whitman Center at Rutgers University, and a critic of the way that economic globalisation is currently working, has stated that 'We are living in a McWorld. We need to globalise democratic institutions in order to keep economic globalisation in check'. Barber has highlighted drugs, pornography and the 'war on children' as the major negative impacts of globalisation. Many see the growing influence of global civil society as a major factor in countering the negative aspects of globalisation. The general message coming from this disparate array of individuals and organisations is that the starting point is to fundamentally change the way in which the global economy is organised.

2 Redesigning the Global Economy

Critics of the way globalisation is proceeding at present argue for a number of significant changes to the global system including the following:

- The establishment of a global central bank.
- A revamping of the IMF to make it more democratic.
- A 'Tobin Tax' on international financial transactions to reduce speculation.
- The establishment of a Global Environmental Organisation to monitor and reduce the impact of economic activity.
- The control of capital for the public good.

The major overall objective is that the two prime movers of the global economy, the economically powerful nation states and transnational corporations, become more accountable to the people of the planet and that all the impacts of economic activity are taken into account in the decision-making process. The goal must be to spread the benefits of globalisation more widely so that all peoples feel included in the global improvement in the quality of life.

3 Sustainable Development

Present levels of consumption are creating an unsustainable demand for many resources. As the world globalises, the effects of excess demand crosses borders and effects societies economically, socially and environmentally. There are two aspects to the problem:

- The impact of developed world consumption on the environment of developing countries.
- The impact of developing countries acquiring developed world consumption habits.

Getting the global community to agree on a common course of action will become even more important in the future with the increased scale of 'spillovers' generated by a more interconnected world and global economy. Sustainable development requires action across many sectors, including water, energy, health, agriculture and biodiversity. As the UN Environment Programme's Geo-2000 report points out, the 'time for a rational, well-planned transition to a sustainable system is running out'.

CASE STUDY: THE CHANGING NATURE OF WORK IN THE UK

The nature of work in the UK has changed markedly over the last 50 years and it will undoubtedly undergo further transformation in the decades to come as the process of globalisation continues. The key questions would seem to be:

1. Will even fewer people work in the primary sector and which tasks will be performed by those that remain?
2. How far will manufacturing employment fall and which products will Britain still produce?
3. Which service sector jobs will decline and which will increase in importance?
4. Which totally new services will begin to provide employment in the future?
5. How many people will be unemployed at various stages in the future and what status and standard of living will they have?
6. What changes will occur in: (a) the working week; (b) paid holidays; (c) retirement age; (d) pensions; (e) the school leaving age; (f) working conditions; (g) the location of employment?
7. What control will the British government have over these issues?

Employment is one of the major concerns of most people, if not the most important overall governing factor in their lives. It is the income derived from employment that influences so many aspects of an individual's quality of life. The notion of security and control is important to most peoples' feeling of well-being and this is usually challenged by rapid change. The Britain of 2025 is likely to be very different from the present state of the country. Today's sixth-formers will be in the prime of their working lives in a world that, in all probability, will be much more globalised than today.

With further advances in ICT there will be a greater opportunity for more people to work from home. Telemobility will allow many people to perform the same tasks from home that they now do in their office. However, a decade ago it was thought that higher technology home working would be more important now than it has actually turned out to be. It seems that the physical clustering of people in organisations has proved more difficult to break down than many commentators thought.

It seems likely that the increases in international commuting within the EU, on both a daily and a weekly basis, will continue, as the real cost of air transport in particular continues to fall. For example, the dynamism of the Irish economy over the last decade has attracted many British workers to fly to Dublin early on Monday morning, stay over during the week, and return to London, Birmingham, Manchester and other cities on Friday evening.

It is also likely that employment migration (geographical mobility) within the EU will increase as economic and psychological barriers to movement recede. The degree of occupational mobility should also increase as the pace of change quickens.

4 Religious Fundamentalism and Globalisation

The increasing pace of globalisation in recent decades has been paralleled by the rise of religious fundamentalism. Many writers see the latter as a direct and indirect reaction to the former. In the West the term 'fundamentalism' invariably means Islamic fundamentalism. The fundamentalist revival in Islam that began in the 1970s is, rightly or wrongly, a major concern to many people and governments in the West. This is because this 'aggressive' faction of Islam largely rejects Western modernisation and secularism. The call is for 'Islamisation' whereby:

- sharia law replaces secular law
- education centres on the Koran
- the economic system is oriented to redistribution rather than the accumulation of wealth by individuals
- cultural products (TV, music, etc.) are tightly controlled.

The irony for many people in the West is that globalisation has made a pan-Islamic movement possible. For many outside the West globalisation and secularism go hand in hand whereby increasing wealth spawns moral decadence.

However, the influence of fundamentalism on other religions should not be underestimated. The development of the New Christian Right in the USA has had a significant influence on attitudes and policy-making in that country.

Summary

- There are various predictions about the eventual outcomes of globalisation.
- Critics of the way globalisation is preceding at present want to see a fundamental redesign of the global economy.
- If globalisation is to succeed in the long term it must embrace sustainable development.

- There is a considerable tension between globalisation and religious fundamentalism, particularly of the Islamic kind.
- As globalisation proceeds it will have more and more of an impact on peoples' lives in the UK and elsewhere.

Questions

1. Discuss the possible consequences of advanced globalisation.
2. Why must globalisation embrace sustainable development to a much greater extent in the future?
3. Debate the tensions between globalisation and religious fundamentalism.

Bibliography

Anon. (2001) *The Ruhrgebiet: Facts and Figures,* Kommunalverband Ruhrgebiet, 2001.

Anon. (2002) The German Problem, *Newsweek,* 30 September 2002.

Barber B (1992), Jihad vs McWorld, *The Atlantic Monthly,* March 1992, Vol. 269, No. 3.

Cramp P (2001), *Economic Development,* Anforme.

Dicken P (2001, 3rd edition), *Global Shift: Transforming the World Economy,* Paul Chapman Publishing.

Dunning J (2000), *Regions, Globalization, and the Knowledge-based Economy,* OUP.

Ellwood W (2001), *The No-nonsense Guide to Globalisation,* New Internationalist Publications.

Gilpin R (2000), *The Challenge of Global Capitalism: the World Economy in the 21st Century,* Princeton University Press.

Guinness P (2002), *Migration,* Hodder & Stoughton.

Guinness P and Nagle G (1999), *Advanced Geography: Concepts and Cases,* Hodder & Stoughton.

Hobsbawm E (1992, 2nd edition), *Nations and Nationalism Since 1780,* Cambridge University Press.

Klein N (2000), *No Logo,* Flamingo.

Legrain P (2002), *Open World. The Truth about Globalisation,* Abacus.

Litvin D (2003), *Empires of Profit: Commerce, Conquest and Corporate Responsibility,* Texere.

Luo Y (2001), *China's Service Sector: A New Battlefield for International Corporation,* Copenhagen Business School Press.

Millstone E and Lang T (2003), *The Atlas of Food,* Earthscan.

Ransom D (2001), *The No-nonsense Guide to Fair Trade,* New Internationalist Publications.

Robertson R (1992), *Globalisation,* Sage.

Rugman A (2000), *The End of Globalization,* Random House Business Books.

Smith MP (2001), *Transnational Urbanism: Locating Globalisation,* Blackwell.

Wallerstein I (1979), *The Capitalist World Economy,* Cambridge University Press.

Waters M (2001, 2nd edition), *Globalization,* Routledge.

Wolf M (2001), Will the Nation State Survive Globalization? *Foreign Affairs,* Jan–Feb 2001, Vol. 80, No. 1.

World Development Report (2003), The World Bank and OUP.

Index